# Ecolanda

## The First Emerald Bioeconomy

**Circular biosystems improve productivity and health while cleaning and restoring our environment.**

**Mark R. Edwards**

**John G. O'Hare**

**SCAD – Southern Cross Agri-energy Development**

## Dedication

*To our future children and their children. May they have the natural
resources and colourful biodiversity they deserve*

## Table of Contents



Mark Edwards, Director SCAD Ecolanda, Biotech and Ecology

John O'Hare, Managing Director, SCAD Queensland and SCAD Ecolanda

**ISBN:** 9798732710540

# Acknowledgements

| | | |
|---|---|---|
| Xuemei Bai | David Punchard | Charles Greene |
| Robert Henrikson | Zhao Yiyuan | Bruce Rittmann |
| Stephen Mayfield | Benjamin Brant | Ann Ewen |
| Paul Bryne | John Benemann | Ben Wells |
| Caroline Hragellund Ruckers | Lieve Laurence | Ira Levine |
| Anastasia O'Rouke | Maura Flight | Klaus Lackner |
| Amha Belay | Mimi Hall | Chris Hall |
| Susan Shultz | Ben Cloud | Efraín Reséndiz |
| Qiang Hu | Emily McKoy | Cathy Stanton |
| Antoine Schellinger | Elham Fini | Mona Molnvik |
| Zuzu Meta | Alice Burns | Andy Ayers |
| Ras Neillsen | Richard Bellingham | Mark Ewen |
| Gary Nicholson | Ken Warneke | Rod Missen |
| Godfrey Dol | Ted Gribble | Lyle Ewanchuk |
| Gordon Lacey | William Lee | Dato Annie |

## SCAD Ecolanda — Executive Summary
### SCAD Designs, Builds and Operates an Emerald Environment with Resilient Economic and Social Development

**Good Work**
- 10,000+ good jobs
- All skill levels
- Excellent training

**Strong Society**
- Health, well-being
- Strong education
- Peace and justices

**Healthy Food**
- Affordable food
- Fresh and local
- Superior nutrition

**Ecosystems**
- Healthy and verdant
- Restore fertility
- Restore fins and shells

Economy

Society

Biosphere

**Social Equity**
- Reduce poverty
- Gender equity
- Reduce hunger

**Strong Infrastructure**
- Smart infrastructure
- Smart, eco-cities
- Affordable energy

**Clean Energy**
- Renewable
- Microgrids + storage
- Batteries + hydrogen

**Biosystems**
- Clean air and water
- Restore eco-health
- Restore biodiversity

Use carbon once, shame on you.

Use carbon 2+, celebrate you.

SCAD cultivates a broad array of macro and microcrop bioproducts such as:

- **Energetic bioproducts** – food, feed, fertilizer, nutraceuticals, cosmeceuticals, pharmaceuticals and biofuels.
- **Environmental restoration** – decarbonization, remediation of air, water and soils as well as ecosystem and biodiversity recovery.
- **Carbon-negative, biodegradable substitutes** – biotextiles, pigments, bioplastics, biofilms, packaging, construction materials, concrete, asphalt and advanced composites.
- **Nutritionals** – foods and medicines that address stunting, wasting, premature births, developmental disabilities, obesity, diabetes, cancers, CHD and brain diseases.

**SCAD Ecolanda** leads the transformation to a robust bioeconomy with clean, fossil-free agri-energy with no waste or pollution. Ecolanda restores health to society and to ecosystems.

We practice **Abundance** and biocycle nutrients with BioRenew to preserve natural resources and to restore health to people, animals and ecosystems. We produce clean food, energy and water with net-zero carbon emissions and a positive eco-footprint.

**Global firsts:**
1. Practice **Abundance** methods with **BioRenew** on a commercial scale.
2. Create the highest crop productivity per hectare – plant and animal.
3. Highest decarbonization — capture and reuse > 50MMT CO2e
4. Use minimal or no extracted fossil resources - land, water, chemicals.
5. Produce cleaner and healthier bioproducts with net-zero waste.
6. Highest nutrient biocycle - carbon, methane and other nutrients.
7. Build a fossil-free agri-energy system with a positive eco-footprint.
8. Cultivate happiness with superior nutrition, health and lifestyles.

## SCAD designs, builds and operates five clean fossil-free sectors:

| Smart Agriculture | Smart Energy | Smart Water | Smart Waste | Smart Ecocity |
|---|---|---|---|---|

**Smart Agriculture**
- **Zero waste**
- BioRenew biocycles waste
- Vertical farms
- Higher production
- Lower costs
- Happier animals

**Smart Energy**
- **100% less fossil fuel**
- Solar, wind, hydro, waves, geothermal
- Advanced production efficiencies
- Advanced storage
- Smart microgrids

**Smart Water**
- **80% less fresh water consumed**
- Brine, waste, ocean, recycled
- Minimize water loss- evaporation
- BioRenew cycles and cleans water

**Smart Waste**
- **100% less landfill**
- Zero GHG
- Near-zero runoff
- Waste to energy
- BioRenew cycles waste and makes new bioproducts

**Smart Ecocity**
- **Zero waste**
- 70% reduction in water & energy
- Everything is connected
- Green transport
- Green space

John O'Hare johare@scadev.net
Mark Edwards medwards@scadev.net

# 1. Emerald Bioeconomy

*When it is obvious that the goals cannot*
*be reached, don't adjust the goals,*
*adjust the action steps.*    *— Confucius*

Ecolanda achieves a sustainable circular bioeconomy due to an architectural design based on nature, smart biotechnology and 7-generation stewardship.

A **bioeconomy** uses renewable bioresources from land and sea. Bioresources include crops, forests, fish, animals and micro-organisms to produce food, materials, consumer products and green energy. Most bioeconomy projects hope to reduce a carbon footprint sometime in the future.

Ecolanda will be the first **emerald bioeconomy**. The architecture connects every system to assure the sustainable cultivation of superior goods with minimal extraction and zero waste or pollution. Ecolanda is built to be carbon neutral on day one.

Ecolanda's emerald bioeconomy goes far beyond carbon neutrality by producing superior bioproducts in an eco-responsible manner that:

1. Are **carbon negative**, capturing and reusing carbon and sequestering carbon.
2. **Integrate smart biosystems** that clean and restore air, water and ecosystems while cultivating valuable bioproducts.
3. Apply **7-generation stewardship**.

Each Ecolanda project plans to capture and reuse over 70 million metric tons of carbon annually.

## 7-generation stewardship.

Ecolanda evolved from 40-years of global agri-energy sustainability research. The principal investigation focused on:

*Will the proposed process operate cleanly*
*and smoothly for 7-generations?*

Indigenous people have much to teach us. Native people lived for centuries in their sustainable communities. Indigenous people revered nature – air, water, land and wildlife.

Native communities often made decisions based on 7-generation stewardship. Long before the current attention to sustainable systems, Native people asked what impact their decision today would have on the welfare and wellbeing of children born in 7-generations, 140 years.

Ecolanda uses 7-generation stewardship to ensure positive impact on nations and communities. Several Ecolanda projects are planned on four continents. Each takes the name of the host Nation or region and selects clean biotechnologies based on 7-generation stewardship.

Idyllic cities and communities have been proposed through history. A few have been built, such as Gaviotas, a mountain village in Colombia, South America.[1] Most ideal communities were based on impractical theory, not reality.

## Ecolanda foundation

Ecolanda benefits from a solid foundation built primarily on tested biosolutions that exist today. Dozens of site visits, hundreds of interviews and many techno-evaluations drove the selection of eco-friendly solutions. Many excellent tools are available today but are not sustainable as stand-alone systems.

## SCAD, Southern Cross Agri-energy Development

SCAD serves as the Ecolanda systems integrator and links clean technologies into circular biosystems. SCAD's analysis considered and tested the pros and cons for mechanical and biotechnologies that claimed agri-energy sustainability.

Several conclusions emerged:

1. Extractive methods are sustainable only until the first needed substance goes extinct or extraction becomes too expensive. Extractive processes like industrial agriculture will leave our future children without the precious natural resources they need.

Prior generations did not concern themselves with resource scarcity because resources were plentiful.

Today, only 70 years since the start of the Green Revolution in agriculture, many of the valuable natural resources needed for food and fibre production are past their peak extraction and are moving rapidly toward extinction. Soil nutrients, fresh water and phosphorus are prime examples.

2. Industrial mechanical methods are not sustainable. Continuous extraction, waste and pollution add expense and destroy ecosystems. Circular biosolutions avoid extraction and waste and are sustainable.

Farmers have gone bankrupt and left their farms in many communities due to weather, water or the increasing cost of agri-inputs. Extraction continually becomes more expensive as the quality of ore or target substance diminishes and refining becomes more difficult.

Systematic nutrient extraction from cropland has left more cropland exhausted and abandoned than is farmed today.[2] Most good cropland has been farmed for decades. Farmers search for new cropland but find the land less flat, less fertile and substantially more difficult to farm. Cropland expansion drives deforestation which contributes 20% of annual GHG to the atmosphere.[3]

3. Eco-friendly methods may not sustainable when they operate alone.

They need the substantial benefits that accrue from systems integration. Integrated systems link each process with upstream elements for inputs and downstream systems for waste capture and nutrient reuse.

4. Industrial mechanical agri-energy solutions are not long-term viable.

Agri-energy solutions need to be designed to resolve the substantial risks from climate change. Agri-production must stop the systemic degradation of our atmosphere, cropland, waterways and ecosystems.

The top five global meat and dairy firms are now responsible for more annual greenhouse gas, GHG, emissions than Exxon, Shell or BP.[4] Industrial agriculture also is responsible for over 80% of hazardous water pollution.

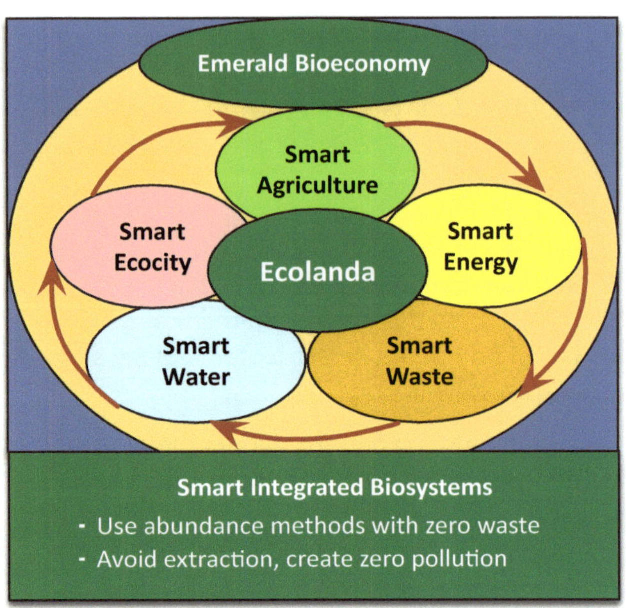



5. Sustainable solutions must not only avoid adding more environmental pollutants but also remediate toxic wastes accumulated from decades of industrial production.

Our children and their children deserve clean air, water, lush ecosystems and oceans filled with rich marine life. Biosolutions to attain these goals are examined in the following chapters.

### National Stewardship

National leaders in the host country shape each Ecolanda project. Ecolanda planning commences with the host country selecting a National Stewardship Team. Ecolanda typically represents the largest private investment in a country's history, €15 Billion. Joint stewardship and collaborative planning develop strong National goals, timelines and action steps.

### Integrated biosolutions

Ecolanda demonstrates how integrated biological systems, circular economies, create significantly higher value compared with industrial mechanical methods. National value metrics report benefits in health, economics, society and environment.

Residents and neighboring communities will embrace Ecolanda due to improvements in health, affordable and plentiful food and living standards.

Ecolanda serves ecotourists, students, teachers and scientists with fascinating demonstrations of integrated sustainable systems. Ecolanda's architecture creates the highest agri-energy productivity while capturing carbon.

## SMART biosystems

SCAD employs **SMART** integrated biosystems. These **S**ustainable **M**icro **A**bundance and **R**egenerative **T**echnologies are each sustainable and non-pollutive for 7-generations.

**Micro Abundance** methods mimic nature and create superior food and other consumer and industrial products in a circular bioeconomy.

Nature has grown microcrops consistently daily for 3.5 billion years. Trillions of hungry consumers eat microcrops daily because they typically deliver over twice the nutrition per bite as land-based crops. Small creatures such as shell and finfish depend on algae for food. Large animals including giant blue whales get their primary nutrition from algae and algae feeders like krill.

Microcrops such as algae provide 40% of the new biomass on earth daily. These microscopic plants produce 70% of the planet's new oxygen daily.

Microcrops enable growers to cultivate a wide array of valuable bioproducts without consuming precious fossil resources. Abundance growing methods provide significant advantages compared with environmentally destructive industrial mechanical agriculture. Abundance methods allow growers to:

- Biocycle waste stream nutrients from air and water, which reduces input costs and avoids natural resource extraction.
- Use minimal or zero cropland, fossil fuels, fresh water, chemical fertilisers, pesticides or other agri-poisons.
- Create zero waste, GHG emissions and pollution. The only emission is pure oxygen.
- Clean air, water and ecosystems while producing valuable food, consumer and industrial bioproducts.

## Regenerative technologies

An end to the enormous extraction, waste and pollution from mechanical agri-energy production provides substantial value. However, nature works slowly and would require centuries to remediate existing environmental contamination.

Ecolanda goes much further than the "Do no harm" command with biosolutions that accelerate nature's ecosystem recovery.

**Regenerative technologies** play a critical role in Ecolanda's 7-generation stewardship. Abundance methods avoid extraction, which ensures that plentiful natural resources will be available for future generations to enjoy.

Regenerative technologies are highly productive as they employ microcrops to produce healthy and clean food, feed and other bioproducts. The abundance cultivation process cleans, repairs and restores ecosystems degraded or destroyed by mines, refineries and industrial production.

Ecolanda's regenerative technologies begin with biocycling carbon and other pollutants from the atmosphere. Every ton of microcrop food, feed or fertilizer captures two tons of $CO_2$. Microcrop cultivation cleans and clears the atmosphere we share. Polluted air bubbles through biosystems where tiny microcrop cells removes contaminants. Extreme contamination may require more than one pass through a biosystem.

Ecolanda's regenerative biosystems can clean polluted water and restore extra fresh blue water for drinking and other community needs.

Industrial production has seriously degraded verdant ecosystems and destroyed their colorful biodiversity. Ecolanda's highly productive smart biosystems repair and restore health to degraded forests, fields and waterways.

Healthy ecosystems are attractive to birds, bees, butterflies and other land and marine creatures. Restored ecosystems allow bright and beautiful biodiversity to repopulate and sustain biodiverse life for many generations.

### *Ecolanda priorities*

Ecolanda growers orchestrate smart biosystems to optimize health, productivity and safety for people, producers and our planet.

Industrial farmers suffer among the highest health, disability and premature death danger of any occupation. Farmers are injured by huge equipment and physical labor and poisoned by dust, black soot, fertilizers and pesticides.

Biosystems assure grower safety. Cultivation requires no heavy physical labor or large tractors or harvesters. Growers are not overwhelmed by dust and are not exposed to hazardous chemical fertilizers or agri-poisons.

Even if technology advances removed all air, water and ecosystem pollution, industrial systems are so inefficient they create far too much waste.

Waste adds expense to every step in the industrial supply chain. Industrial agriculture typically wastes more than 50% of all the inputs farmers apply to their crops. Those wastes massively pollute our air, water, ecosystems and oceans.

Ecolanda operates efficiently across five highly integrated vertical sectors: agri-energy-waste-water and a smart ecocity.

Ecolanda will create the **highest global:**

1. Sustainable food yield per hectare
2. Nutrient density and diversity per bite
3. CCU, carbon capture and utilization, reuse
4. Sustainability metrics & smart biosolutions
5. Renewable energy generation, distribution, storage and use efficiencies

Goals include the **lowest global**:

1. Carbon, water and ecological footprint
2. Net food and bioproduct production cost
3. Natural resource extraction, waste, emissions and pollution
4. Cropland, fossil fuel and water use
5. Chemical fertilizer and pesticide use

A National Task Force selects additional National priorities.

### *National priorities*

The SCAD Ecolanda team co-plans each Ecolanda project priorities with the National Task Force. Most countries have National Priorities for health, economy, society and environment.

The two teams jointly create a set of goals, timelines and accountabilities in each area. Example goals are:

**Health**

- Reduce malnutrition 80%.
- Reduce preterm newborn deliveries 60%
- Reduce stunting and wasting 50%.
- Improve health and vitality 30%.

**Economy**

- Increase employment 30%.
- Improve rural economies 20%.
- Increase exports 20%.
- Upgrade clean-tech innovation and entrepreneurship 40%.

**Society**

- Increase life-quality and happiness 30%.
- Enhance key education metrics 50%.
- Improve social satisfaction 30%.
- Improve community engagement 50%.

**Environment**

- Lead the world in decarbonization.
- Slow deforestation 75%.
- Reclaim 20,000 hectors of degraded crop and forest land.
- Demonstrate 10 smart cleantech biosystems that pay for themselves while they repair and restore ecosystems.

National goals change with experience and demonstration of successful novel biosolutions. The SCAD Ecolanda team and the National Team will prepare a comprehensive National Goal progress report annually.

Models and projections play an important role in Ecolanda metrics. Data on goal progress allow the National Team and institutional experts to model smart systems. These models help the National Team make policy decisions the benefit communities and the nation.

### Respect

SCAD's primary purpose focuses on lifting life quality for people, producers and ecosystems.

SCAD respects People and ensures that People there are healthier, happier, better educated and better employed then when SCAD arrived.

### Transparency

*In many areas of human endeavor, especially health, economics, society and environment, superior decisions depend on accurate metrics.*

Ecolanda success depends on robust metrics. Mechanical industry uses general measures to assure industrial production stays within certain parameters.

Biosystems are both more complex than mechanical methods and much smaller. Nano and microorganisms are usually not visible without a microscope. SCAD's smart systems use 5 to 10 times more metrics than mechanical analogs.

Ecolanda's smart systems operate with sensors and monitors that provide big data 24/7. All inputs and outputs are measured in order to optimize operations and scout for improvements.

Live metrics flow to dashboards that assemble the data in templates that are easy to read. Key people on the National Team or their designees have access to live dashboards.

Selected experts at National or State Academies and Universities will have access to these reports. The reports and data behind the reports will allow external experts to analyze and validate progress.

Dashboards roll up into daily, weekly and monthly reports. These rollups provide quick biosystem efficiency summaries. Reports help spot inefficiencies or choke points when they occur.

SCAD has created strong alliances with world-class scientists globally that will have access to selected data and reports. These subject matter experts will provide third-party validation for progress on goals and objectives in each area.

### Critical export

The most important Ecolanda export will be circular econometrics. The metrics compare traditional industrial production with biosystems.

The first question most people ask is: "If biosystems are superior in so many ways to mechanical systems, why do investors continue to select industrial methods?"

The metrics for proof do not exist today. No large integrated circular biosystems currently exist. Investors prefer to put their money in models that generate an immediate and known return. Investments continue even if the processes are viable only short term and continuously impose havoc on our environment.

Ecolanda will make the critical eco-metrics available free globally. Metrics will allow investors clean options when deciding where to make investments.

The second argument against clean investment is: "It costs too much." Eco-metrics will show clearly that communities and nations are paying a far higher price today for dirty industrial production. They can shift their existing investments to cleantech and save money and their environment while creating good new jobs.

### Site selection

SCAD uses a comprehensive matrix for site selection. Selection begins with a site visit to study the tentative location and determine if the area fits Ecolanda development parameters. Site selection examines factors supporting Ecolanda's integrated systems model. Criteria analyze:

1. Geography – topography, latitude, land configuration, and composition.
2. Climate – number of sunny days, high and low temperatures, humidity and other climatic factors.
3. Political – National, regional and local support, commitment to the environment and willingness for permitting.
4. Carbon – carbon trading market available.

5. Infrastructure – roads, communications, internet access, railroads, harbour, airport, schools and medical facilities.

6. Environment – potential for solar, wind and other forms of locally available renewable energy,

7. Land – size, type and quality. SCAD can often use cropland too degraded or worn out for industrial development.

8. Natural resources – water sources and quality, animals and biodiversity.

9. Labour – training, experience, labour relations and willingness to learn new methods.

10. Skilled people – access to scientists, engineers and technologists.

11. Education – schools at all levels, National or State Colleges and Universities

12. Markets – access for Ecolanda inputs and sale of bioproducts.

Site selection allows SCAD leadership to determine the most appropriate location.

### *Finance*

Ecolanda becomes a project of **"national interest"** due to its importance and social and economic scope. It is expected to be the largest private investment in a single project in the country's history. Ecolanda will bring economic gains of more than €200 billion within ten years.

Project financing varies by country. Ecolanda projects are so large that national leaders need to be highly engaged in the process. Extensive coordination between national leadership and SCAD are critical to build Ecolanda successfully.

The public part of the public-private partnership begins with National Leaders making several commitments:

1. Appoint a strong interdisciplinary National Task Force to coordinate Ecolanda designs, plans, budgets and timelines with SCAD.

2. Issue a government bond for €1 – 2 billion (TBA based on Project Scale) to a top 10 bank.

3. Designate minimum two land parcels of 10,000+ hectares, each for a 99-year lease.

4. Agree to appropriate in kind infrastructure improvements and tax incentives

5. SCAD site to be designated an Economic Trade Free Zone for 30 years sustaining the legal conditions for investment.

6. The government's decisions contribute to project competitiveness and reduces construction costs.

The Economic Trade Free Zone will:

- Boost exports and foreign exchange.
- Add value to raw materials produced in the country such as food, feed, biochemicals, biopharmaceuticals, biofuels and bioenergy.
- Create jobs, training and a wide array of advanced technologies.

SCAD orchestrates the private sector part of the partnership with aligned actions:

1. Appoint a strong SCAD interdisciplinary task force for coordination with national, regional and local entities via a National Task Force.

2. Commit to a €15+ billion private investment to build Ecolanda in the country.

3. Present architectural designs, plans, timelines and accountabilities.

4. Share regular progress reports on a mutually approved schedule.

SCAD will communicate with National Task Force on investment timing. SCAD creates a broad project strategy with specific plans. Timelines on plans may change based on mutual agreement among the relevant parties.

The next section highlights how Ecolanda uses Abundance agricultural methods to accomplish both high productivity and ecological goals.

# 2. SCAD Ecolanda

*Vision: SCAD Ecolanda leads the transformation to clean, fossil-free agri-energy with zero waste or pollution and restores health to people, producers and our environment.*

Societies globally are systematically running out of natural resources. Scarcity constantly drives-up manufacturing cost and consumer prices. Every minute, industrial production extracts more irreplaceable natural resources to feed its insatiable appetite.

Industrial systems accelerate fossil resource loss, overconsumption, waste and pollution. Cropland has become so scarce that farmers are clear-cutting ancient forests to grow animal feed.

All food-growing continents face water scarcity. Entire cropland regions are abandoned as wells and surface water sources go dry. Fossil fuels continually pollute our air with $CO_2$, poisonous nitric oxides and deadly black soot particulates.

Chemical fertilisers erode from cropland and have created over 400 dead zones globally in waterways and oceans where eutrophication has suffocated all life. Deadly pesticides kill good and bad insects and exterminate biodiversity.

When our children need natural resources, they will be gone. Industrial emissions continually accelerate global climate chaos. Mechanical systems systemically degrade and destroy ecosystems and their colourful biodiversity.

Population growth drives industrial production in the wrong direction, leading to more overconsumption, GHG emissions and ecological contamination.

Our children deserve a change in direction.

## Ecolanda Strategy: Abundance

The world needs a new paradigm that produces superior food and other consumer goods faster and sustainably. New methods should assist natural systems with accelerated environmental restoration and biodiversity recovery.

Ecolanda addresses each of these critical issues by transforming industrial mechanical production to biological solutions using **abundance** methods.

Ecolanda transforms industrial food and consumer products creation to abundance methods that avoid natural resource extraction, overconsumption, waste and pollution. Abundance cultivation may occur on non-cropland. This saves valuable ecosystems and their biodiversity for future generations.

## Circular Bioeconomy

Abundance methods enable the creation of a 360bioeconomy. Circular biosystems provide the most sustainable and ecologically responsible large-scale development. Biosystems maximize health, economic and social goals while repairing and restoring the environment. Ecolanda uses the circular Nrich process to enhance nutrition in a manner that delivers healthier food for people, animals and plants.

Ecolanda uses the **BioRenew** process to biocycle carbon and other valuable nutrients. Ecolanda creates net-zero emissions while capturing, recycling and reusing over 70 million metric tons of $CO_2$ equivalents annually.

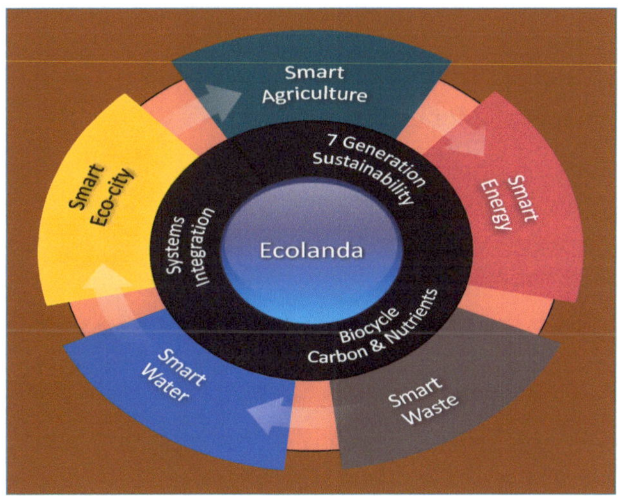

## Decarbonize

Removing carbon from the atmosphere improves air quality by eliminating GHG and other toxic chemicals from industrial stacks. Improved air quality improves health, reduces respiratory diseases and enhances life quality.

Most communities need more blue drinking water. SCAD's BioRenew process creates millions of liters of blue fresh water annually from agricultural, industrial and residential waste streams. Ecolanda's goal: create 20% more fresh blue water than the project uses.

SCAD's Nrich process restores nutrients and fertility to damaged and degraded croplands. Soils often suffer from nutrient depletion from years of farming. Repeated crop production also extracts micronutrients and humus, the critical organic matter that holds moisture and nutrients.

Nrich can restore macro and micronutrients as well as humus to cropland. Nrich can clean and restore ecosystems. These integrated systems allow biodiversity to recover.

Ecolanda creates over 10,000 direct jobs and more than 50,000 indirect jobs. These jobs provide sustainable, long-term benefits to society and lift the economy and living standards.

Ecolanda cultivates affordable food, feed, fibre and a wide array of other valuable bioproducts. The agri-energy sector produces health and affordable food for over 5 million people.

Sustainable food cultivation ensures continuity and improved quality for the region and for the nation's food supply. Ecolanda also cultivates food, feed, fibre and renewable energy for local markets and for export.

SCAD (Southern Cross Agri-Energy Development), based in Australia, designs, builds and operates the world's most productive natural fossil-free bioeconomies while cleaning the environment.

SCAD Ecolanda projects are planned for several countries. The description here focuses on what Ecolanda bioeconomies can do and strategies for effective bioeconomy design, construction and operation.

## Ecolanda strategies

The Ecolanda architecture applies Abundance, strategies to address substantial economic and ecological challenges. These approaches provide a 180° disruptive change from modern industrial mechanical production.

| Ecolanda — BLINC  Strategic Differences | |
| --- | --- |
| **Industrial Production** | **Abundance Methods** |
| 1. Mechanical devices | 1. Biological systems |
| 2. Fossil energy | 2. Light energy |
| 3. Independent systems | 3. Integrated systems |
| 4. Land-based crops | 4. Nano and microcrops |
| 5. Linear economy | 5. Circular bioeconomy |

Ecolanda strategies apply Abundance methods and integrate across sectors to:

1. **Stop systemic damage** inflicted by industrial methods – extraction, overconsumption. waste, emissions and pollution.
2. **Repair ecosystem damage** caused by destructive industrial processes and allows the restoration of biodiversity.

Industrial agriculture offers sharp contrast to abundance methods. Industrial manufacturing does similar damage for other consumer products.

## Mechanical agriculture

Industrial mechanical agriculture drives huge fossil-fuel machinery designed to strip broad expanses of natural ecosystems. Machinery consumes massive fossil resources in order to force nature to do the farmer's will.

Mechanical methods systematically abuse cropland with tractors, ploughs, disks, rippers, compactors, harvesters, fertilisers, pesticides and erosion until the living soil dies from exhaustion.

After years of abusing soils, farmers abandon destroyed cropland. They leave millions of once pristine cropland hectares degraded, destroyed and unavailable for future generations. More farmland has been abandoned globally in the last 60 years than is farmed today.[5]

## *Industrial mechanical agriculture*

Farmers employ heavy mechanical equipment to:

1. Carve out nature - scrape the natural ecosystem "clean" to make way for GMO monocultures.
2. Slash, gash and rip the topsoil deeply to prepare the soil bed.
3. Destroy biodiversity.
4. Disk the topsoil flat to prepare for planting.
5. Crush and compact the soil with monstrous heavy equipment.
6. Degrade living soil with herbicides and cultivation.
7. Crush soils with irrigation and wet-ground compaction.
8. Kill microorganisms with tons of fertiliser, herbicides, fungicides and pesticides.
9. Erode topsoil, fertilisers and pesticides that migrate into waterways and rural ecosystems.
10. Harvest once a year.
11. Extract huge stores of macro, micro-nutrients and humus with every crop, without replacement.

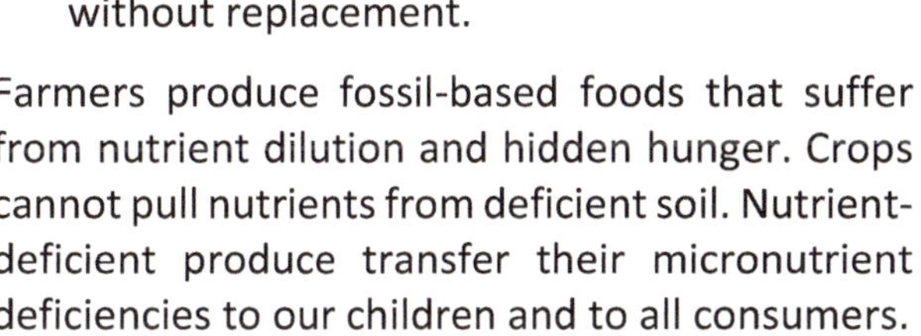

Farmers produce fossil-based foods that suffer from nutrient dilution and hidden hunger. Crops cannot pull nutrients from deficient soil. Nutrient-deficient produce transfer their micronutrient deficiencies to our children and to all consumers.

## *Biological – Abundance methods*

Biological systems extract nothing. They do not overconsume natural resources or create waste. Nature's integrated sustainable systems do not pollute air, water or soil. Nature's circular system works exquisitely with the cycle of life.

Nature nurtures food cultures gently without mechanical intrusion. Nature has used biological systems to grow nutritious food for trillions of consumers daily for over 2 billion years.

Ecolanda growers use abundance biological methods that work in harmony with nature to cultivate healthier, truly sustainable freedom foods with microcrops grown with free energy – sunshine.

Abundance methods allow nature to work gently yet highly effectively. Microcrops grow 20 - 50 times faster than fossil-fuel crops.

1. Build a raceway or indoor biosystem and cultivate microcrops. No cropland needed.
2. Energy – gravity and sunshine. No fossil fuels.
3. Zero fresh water, brine, waste or ocean water.
4. Zero chemical fertilisers
5. Zero pesticides, herbicides or fungicides.
6. Restores ecosystems – air, water, and biodiversity, flora and fauna.
7. Harvest daily, about 320 days a year.

Abundance growers cultivate freedom foods that are healthier for people, producers, animals and our planet. Freedom foods may be made in any form, shape, taste or color.

Nature continually cleans and restores health to air, water, soil and ecosystems. Natural systems require hundreds of years to repair damage from only a few decades of industrial production.

Ecolanda offers several paths to accelerate nature's ability to repair ecosystems and restore biodiversity with novel biotechnologies.

## Light energy

Abundance growers use light, solar energy and photosynthesis as the engine of choice in place of fossil fuels to biocycle nutrients. Light energy may be augmented with concentrated light forms such as photovoltaic and concentrated solar.

Renewable solar and wind energy drive integrated processes across the five smart sectors.

Instead of extracting mined chemical fertilisers, growers use the natural photosynthetic-driven BioRenew process to capture, recycle and repurpose carbon and other nutrients from waste streams.

Reusing nutrients efficiently saves growers money and avoids massive emissions and pollution.

Ecolanda farmers cultivate food in sustainable bio-systems that mitigate climate change as they assimilate rather than emit carbon.

The only emission from a biosystems is pure oxygen. Biosystems enhance social and economic growth free of resource over-consumption and pollution.

Abundance methods flip food cultivation from a major GHG emitter, about 26% of global emissions, to a negative carbon process. Each Ecolanda project plans to capture over 70 million metric tons of $CO_2$ equivalents annually.

## Integrated systems

Ecolanda differs from other agri and economic development because the architecture mimics nature. Each area – agri, energy, waste, water and the ecocity – integrate in a symbiotic manner. The natural architecture allows substantial resource savings compared with modern industrial agriculture and development processes.

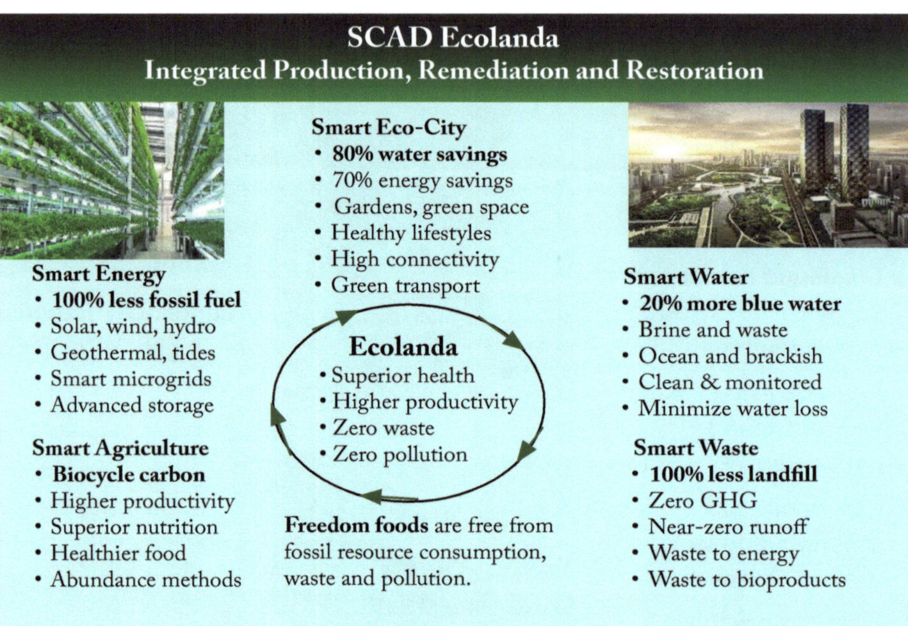



Sustainable agri-energy requires extensive energy, water and waste management. SCAD applies a suite of renewable biological and energy technologies to sustainably manage these critical resources.

Traditional agriculture and animal production are well-known massive emitters of severe pollutants to air, water and soils. SCAD technologies mitigate those pollutants and pursue a goal of net-zero air, water and soil pollution. Ecolanda ensures that no wastes are burned in the open air or buried in waste dumps.

Mechanical agri-energy methods act alone and each waste and emit their own carbon and other pollutants. Each Ecolanda abundance system connects with others to ensure waste from each activity are captured and reused by paired processes.

## Smart energy

Smart energy supplies each sector with renewable energy. Agri-systems both use and produce waste-to-energy and biofuels. Waste-to-energy systems create additional energy in the form of syngas, green hydrogen and other energy products for local use and for export. The smart ecocity produces over 80% of its own energy and food in the municipality.

Smart water integrates across sectors. Industrial agri-systems use massive amounts of blue water. Salt in soil or brine water kills land-based plants due to a plumbing problem. The large salt ions clog the roots, which stops nutrient circulation and the plant withers and dies from lack of nutrients.

Microcrops have no roots and flourish in brine or other non-potable water, including sea water. Most communities have large amounts of wastewater. Smart water treatment and microcrop biosystems allow continual water treatment. Water may be reused many times.

## Nano and microcrops

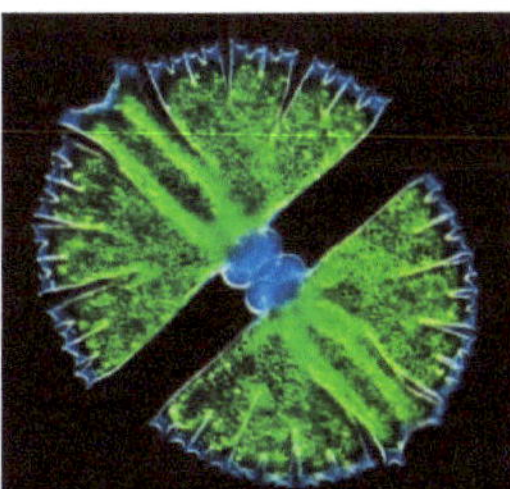

Nano and microcrops are tiny, often just five microns small. These single and multicellular plants without roots grow in all types of water.

Algae are small but powerful. Algae produce 70% of the new oxygen on earth daily, more than twice the oxygen from all the forests and fields combined.

Algae produce 40% of the new biomass on the planet daily. Trillions of consumers eat algae because algae are the most nutritious food.

Tiny microcrops have a huge surface area. A cube of algae the size of a postage stamp, (2 cm x 3 cm) contains more surface area than three football fields. Each algae cell operates independently. The large surface area allows extremely fast assimilation of nutrients in each cell.

Algae grow biomass 50 times faster than terrestrial crops such as rice or corn. A healthy nano-culture can double its biomass each day.

Ecolanda's microcrops are nurtured gently without large mechanical tractors or trucks. Some biosystems use paddle wheels while others mix cultures by bubbling in air and $CO_2$. Growing microcrops does not require ripping the land with ploughs and disks. Growers do not have to clear the land and destroy biodiversity.

Industrial farmers often drive big heavy tractors across their cropland seven or more times a season cultivating, thinning, tending to weed invasion, fertilization, applying pesticides and herbicides and harvesting. Nanocrops require none of these mechanical processes, which saves considerable energy, pollution and costs.

A corn or maize farmer uses mechanical devices to add expensive inputs to the crop that requires a full growing season, 120 days, to produce the first gram of food. A 4-meter maize stalk supports a big biomass but contains less than 3% food. Only the kernels on the single cobb are edible and contain just 33% of an incomplete protein.

Industrial farmers must pay for the inputs to produce enormous waste biomass and then pay again for waste disposal. Most economist would question the sensibility of producing food if 97% of the biomass produces only waste.

Microcrops are harvested daily or several times a week, year-round. A microcrop may contain 65% edible protein in every kilo of biomass. The residual biomass contains additional edible and valuable components – vitamins, minerals, oils, lipids, carbohydrates and other compounds. Microcrops may contain 9% ash, which also contains residual nutrients. Ash residues are biocycled in subsequent cultures.

Ecolanda's freshwater savings, 100% compared with industrial farms, comes from growing microcrops for food, feed, biofertiliser, medicines and other bioproducts in non-potable water. Ecolanda growers cultivate an extensive array of microcrops, including algae with net-zero blue water.

*Microcrops cultivated in outdoor raceways*

Water substitution saves blue drinking water. Blue water contains less than 1,000 parts per million, (ppm) saline. Brine salinity, (NaCl) measures higher than blue water and can be 10 times saltier than ocean water, which contains 36,000 ppm salt.

Most communities can find brine water available since half the water stored in the earth is brine.[6] In many places, brine aquifers are so close to surface, water can be removed by a foot pump.

A nutraceutical project in Arizona uses brine water in vertical column photobioreactors to cultivate algae with omega-3 fatty acids, (below).

The project pumps brine water from an aquifer under the Painted Desert left by an ancient ocean.

The highly saline water is used for nine algae cultivation cycles until algae removes most the nutrients. After nutrient extraction, the still salty water flows to a salt lagoon where the water evaporates and the salt recovered.

## Linear industrial production

Industrial firms today follow an inefficient linear production model that imposes high operational costs, relies on increasingly expensive fossil resources and levies an appalling toll on human and producer health and the environment.

Intensive mechanical agriculture has provided food and other goods for the recent 70 years. Modern consumer products come at the extreme cost of extracting and over-consuming natural resources that will be gone when they are needed by our children and their children.

Mechanical methods require huge amounts of fossil energy because fossil fuels power tractors, trucks, manufacturing and other systems.

Wasteful industrial methods systemically make our planet less livable. Industrial waste pollutes and degrades our air with GHG, smog and black soot particulates.

Mechanical agriculture imposes extreme waste to produce low energetic crops like maize where the waste biomass to food ratio is 97% waste, 3% food. Half the agri-inputs used to grow crops go to waste too, including freshwater, fertilizer and pesticides. These wastes pack waterways with eroded toxic chemicals and deadly pesticides.

Croplands have been severely eroded and exhausted. Large cropland regions have been abandoned because soils have been stripped of nutrients or wells have gone dry.

Linear production causes these vital resources to be extracted and transported to farms where they are stored and then used very inefficiently only once. Each year, farmers must repeat the linear process of buy, store, apply to fields and reorder for next year. Farmers have little control over the substantial cost of waste because field crops use inputs inefficiently. Nearly all industrial agricultural input costs are rising due to increasing scarcity.

Waste, which often exceed the weight of the food crop, are often burned or buried. Crop input wastes are carried away in eroding soils, creating extensive air, water and land pollution.

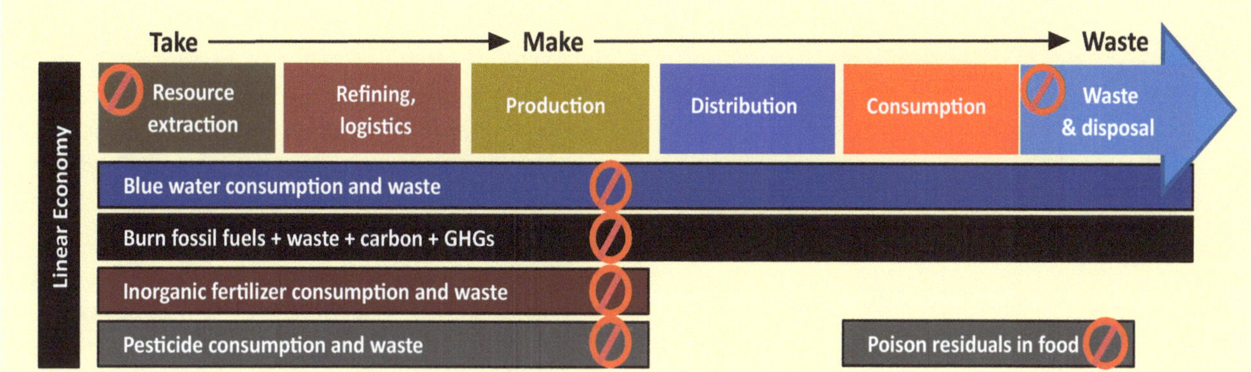



The linear economy model illustrates the extreme waste caused by industrial production. Modern agricultural wastes – cropland, fresh water, fossil fuels, inorganic fertilisers, pesticides – pollute, degrade and poison air, rivers, lakes and oceans.

Crop residues, animal dung and other wastes are often burned creating huge toxic plumes filled with GHG, smog, nitric and sulfur oxides and deadly black soot particulates.

Poor air quality results in early onset of respiratory diseases such as asthma. Prolonged exposure results in respiratory failure and premature death. Oceans are polluted with so much industrial waste that many coastal communities cannot farm fish in the sea. Some seas have more plastic than fish.

A comprehensive study, *The Water Footprint of Humanity,* found that industrial agriculture consumes an incredible 92% of available fresh water.[7] Surface irrigation loses over 50% of the blue water to evaporation before it reaches crops.

Less than half of the chemical fertilisers applied to croplands are assimilated by the plants. The remainder erode on wind and rain and invade local ecosystems. Only 1% of pesticides are assimilated by the crop, leaving 99% to invade and poison local waterways and groundwater.

Numerous scientific studies have shown that dangerous poison residuals remain on produce when they find their way into consumers' homes.[8]

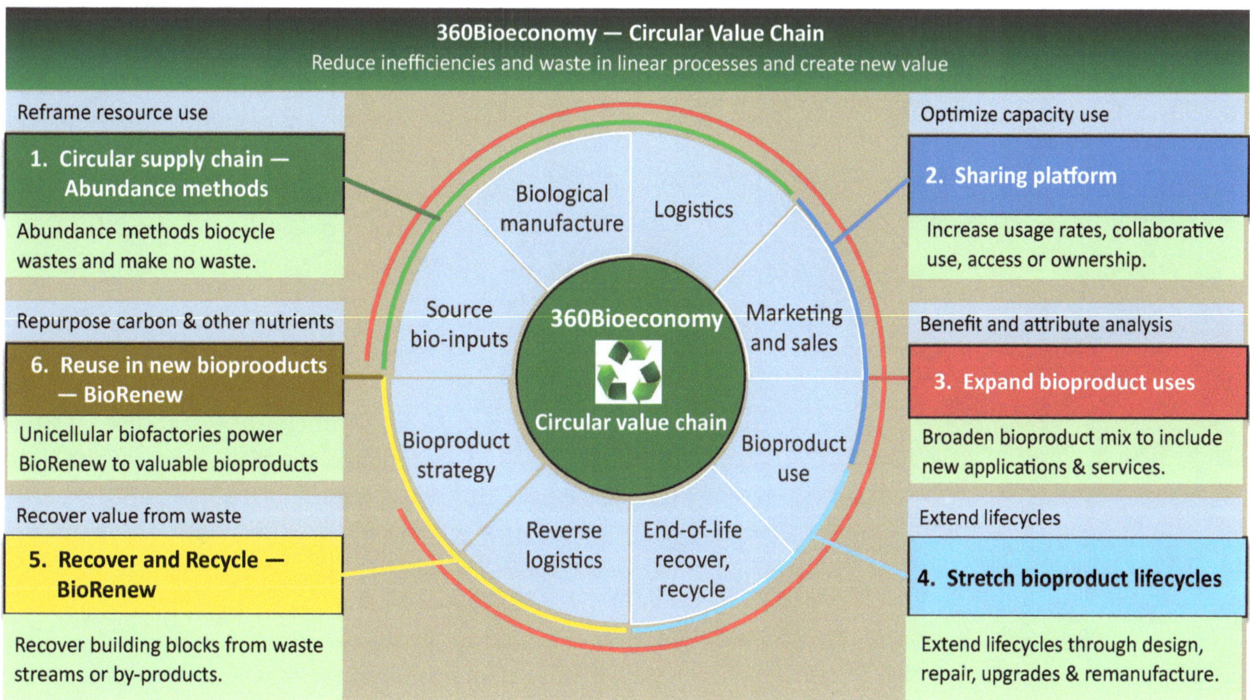

## Ecolanda circular value chain

The significant difference between Ecolanda and other economic development is SCAD's advanced architecture for a circular bioeconomy.

Ecolanda uses a circular value chain that overcomes linear economy problems. SCAD avoids natural resource extraction with BioRenew, which biocycles carbon and other nutrients from waste streams.

The inner circle in the circular diagram shows SCAD's key organizing divisions for the circular value chain. These units integrate across agri-energy-waste-water and smart ecocity.

A 360Bioeconomy represents a sharing platform for biomanufacture. Collaborative use of systems and technologies improves system integration and process efficiencies.

Industrial manufacturing typically creates excessive waste due to the design for a single product. The bioeconomy model changes to expand bioproduct uses and users. Many bioproduct users are local, which reduces costs.

Essentially any product made with industrial manufacture can be made with the SCAD 360Bioeconomy. The ecocity innovation park prioritizes biomanufacture of consumer and industrial products based on social and economic factors. The graphic illustrates the wide array of valuable bioproducts.

## Bioproducts

Bioproducts share important attributes. They are all made with biological processes, so they are biodegradable. Microcrop construction materials supply equal or superior features to industrial versions. When bioconstruction materials come to the end of their useful life, possibly 50 years, they are recovered and recycled. Target compounds are repurposed in new products.

Bioproducts can be made using zero fossil resources, which makes them less expensive and eco-friendly. Many bioproducts such as asphalt and cement substitutes sequester vast amounts of carbon and other polluting substances for generations.

Bioplastics, films, packaging and foams provide valuable substitutes for industrial products that currently are sent to waste dumps and flow into waterways and oceans. Bioproducts break down quickly and do not create land or ocean litter.

A maize farmer grows and sells only maize, the tiny kernels on the cobb as a single commodity product. Microcrops are cultivated with a plan for multiple products from the single culture. Growers tend to avoid commodity products because specialy bioproducts bring more value.

Processing algae biomass may remove 20% of the oil for an Omega-3 fatty acid and the remainder of the oil for cooking. The protein, which may be 65% of the biomass, may be sold as animal feed. The carbohydrates and fibres may be used for bio-construction materials or ruffage for animal feed.

Fossil fuels, petroleum, coal and shale are compsed of fossilized algae from ancient oceans. Nature fossilized the algae deep in the earth with tremendous heat and pressure over 400 million years. Algae biofuels skip those steps. Algae growers can cultivate jet-fuel in a few weeks and produce more fuel weekly. Algae biofuels burn cleanly without smoke or black carbon particulates.

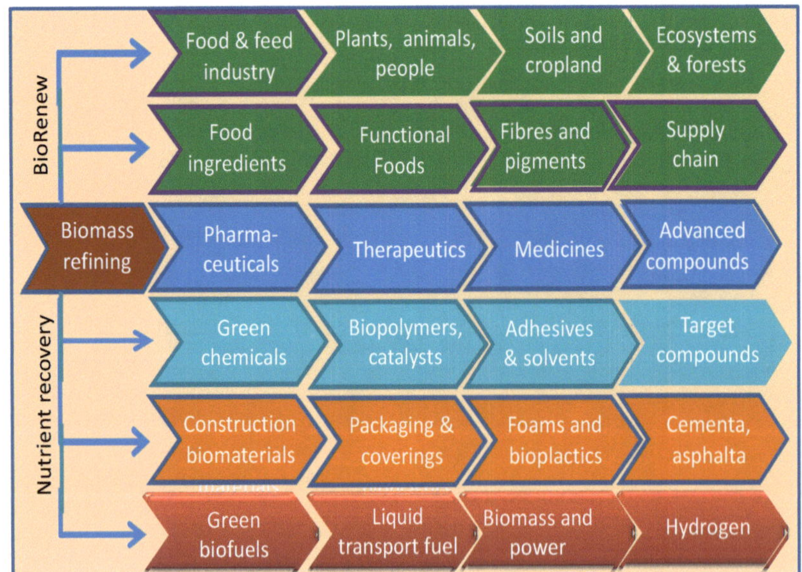

# 3.  Ecolanda Emerald Bioeconomy

*Humans like to consider everything as
linear, when in reality everything cycles.*

Industrial production commonly uses the strip-mining model. Deforestation strips the land of trees and all flora and fauna. Deforestation contributes about 20% of GHG annually to our atmosphere.[9]

Deforestation occurs to clear forests for cropland. The land is stripped of all vegetation in order to avoid weed competition. GMO crops such as soybeans are weak and do not compete successfully with weeds. Consequently, farmers make extensive use of herbicides and other poisons, which systemically kill macro and micro biodiversity in order to grow their crops.

Forests cleared for farming are cultivated for planting in the spring, just in time for high winds and rain. Rains create mudslides, eroding the topsoil and sending debris and agri-chemicals into waterways and valleys.

After only a few years, many forest lands cleared for farms are no longer productive as topsoil has eroded. Farmers move to the next forest and repeat the process. Some countries report that over half the deforested land has been abandoned due to fertility loss.

Intensive mechanical agriculture imposes huge costs of farmers, society and biology. Farming imposes substantial risks, which causes millions of farmer bankruptcies annually. Society loses because mechanical methods consume too many non-renewable resources and discard far too much pollution to air, water and ecosystems. Everyone loses from biodiversity extinction

Intensive mechanical agriculture cannot be redesigned from their costly and inefficient linear production model. The alternative circular model offers far more efficiencies.

## *Circular economy*

Ecolanda models nature and acts gently with beautiful simplicity. Natural resources are neither extracted nor wasted. Nature wastes nothing. Every living organism lives a life. At the end of life, it provides the nutrients for the next cycle of life.

Ecolanda replaces mechanical industrial production with biological solutions that avoid natural resource extraction, waste and pollution. The integrated bioeconomy, (agri-energy-waste-water), process creates an emerald bioeconomy.

The Ecolanda design team develops specific goals in collaboration with stakeholders, including indigenous, local, regional and national residents.

Ecolanda's social and environmental goals are summarized in the following table.

| Ecolanda Social, Economic and Environmental Goals | |
| --- | --- |
| **Social and economic** | **Environment** |
| 1. Lift the standard of living for all | 6. Capture >50 MMT CO2e / year |
| 2. Improve health, vitality and lifestyles | 7. Make >20% net fresh blue water |
| 3. Create new jobs with skills training | 8. Minimize natural resource extraction |
| 4. Craft new materials for housing | 9. Eliminate waste in air, water, soil |
| 5. Engage citizens equitably in decisions | 10. Restore ecosystems and biodiversity |

Social and economic goals focus on improving lives for Ecolanda residents the neighboring communities. A strong emphasis on jobs, education and affordable housing assures improved living standards.

Waste management plays a critical role. All forms of waste are imported by rail from a 300-kilometer radius. Municipal, industrial, medical and hazardous wastes are imported from cities and rural areas and converted to biofuels.

Ecolanda manages all the waste from the agri-energy sector and the smart ecocity. Wastes are biocycled to recover the carbon and other nutrients or transformed into energy products.

Waste management systems typically emit huge amounts of $CO_2$. Ecolanda's $CO_2$ flows into microcrop biosystems where micro-organisms perform their magic to make nutrient-rich biomass that may be used as nutrient-dense animal, fish or plant food.

## Biosolutions

Biosolutions decarbonize ecosystems, capturing and reusing millions of metric tons of carbon and other valuable nutrients, especially phosphorus.

Ecolanda transforms the region into a carbon-negative state while creating the highest yield per hectare for both plants and animals. The agri-energy farm as well as the ecocity operate productively while they clean, enhance and restore degraded ecosystems and accelerate the return of biodiversity.

The smart ecocity goals are summarized in the table. Ecolanda provides significant benefits to both the 10,000 employed and their families. Over 50,000 indirect jobs will also benefit the community.

Citizens benefit from a smart ecocity with affordable housing, schools, retail, green space, clean transportation, medical facilities and an entrepreneurial park. The sustainable ecocity maximizes life quality for its diverse citizens.

| Ecolanda Smart Ecocity and Technology Goals | |
|---|---|
| **Smart Ecocity** | **Technology** |
| 11. Design and build a smart city | 16. Invent the fusion of nature & biotech |
| 12. Maximize happiness and lifestyles | 17. Apply intelligence to agri-energy |
| 13. Create an eco-park for innovation | 18. BioRenew turns trash to cash |
| 14. Leverage distance learning | 19. Smart sensors support production |
| 15. Assure shared community values | 20. Produce biodegradable bioproducts |

It produces 80% of its food locally, 80% of its energy, cleans 90% of its water and creates zero emissions and zero pollution.

Ecolanda's ten-year eco-goals include:

- **10 Billion metric tons of carbon** capture and repurpose along with other GHG
- **100 Billion liters of fresh blue water**
- **10 Billion Wings and Beaks** — restore biodiversity, especially birds, bats and butterflies.
- **10 Billion Fins and Shells** — restore biodiverse marine life.

## Five pillars of sustainable development

Ecolanda's secret to building a bioeconomy relies on integrating the five pillars of sustainable development.

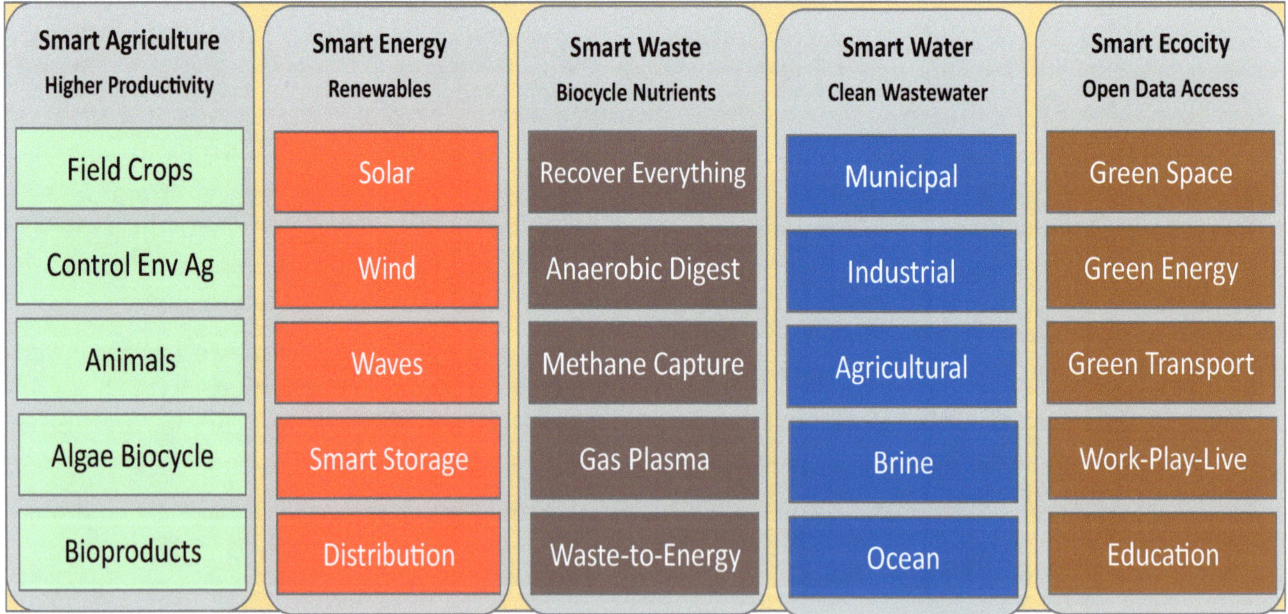

Food, feed, fiber and fertiliser cultivation requires considerable energy generated by renewable sources – wind, solar, waves and geothermal.

SCAD uses several clean waste-to-energy technologies and grows energetic crops for conversion to biofuel. Waste-to-energy processes and biofuel production benefit from reduced costs due to carefully designed carbon-negative architecture.

Water represents a critical variable in SCAD production. Agri-production makes extensive use of non-potable water, including waste, brine, brackish and ocean in addition to blue water. SCAD plans to create 20% more fresh water than the farm and ecocity use. Surplus blue water flows back to the community.

Every Ecolanda sector creates and manages substantial waste. Historically, wastes have been buried in landfills, burned or allowed to invade the local ecology. Waste are used forward in the Ecolanda model. Systems are designed for integration so that any surplus from one process supplies the ingredients for future bioproducts.

Agri wastes flow to waste-to-energy systems that produce syngas, green hydrogen and ammonia. Agri water flows through biosystems to clean water for reuse. Agri cultivation provides food and textiles for the ecocity while agri waste provides clean energy.

The smart ecocity integrates all waste streams. GHG produced by homes, recreation, retail or industry flow by design into algae biosystems that transform the gasses into rich green biomass. Other biosystems clean ecocity water for reuse and for green space and urban garden irrigation.

The ecocity cultivates 80% of its fresh and local food needed by residents and their pets. Communities share urban gardens for fresh produce.

Some restaurants offer green bars where customers select foods growing inside the restaurant. The smart city provides a work-play-living environment with substantial green space.

## Education

Sustainable development requires education and skills training. Education plays a major role across Ecolanda sectors.

SCAD will build a pre-school, grade school, middle and high school. SCAD will work with local school administrators to create appropriate curriculums for all school levels.

SCAD will join a public-private partnership with the appropriate government entities to build and help staff a cleantech college. This cleantech school will train students for jobs in Ecolanda and other cleantech businesses.

Students matriculating from the cleantech college will have the opportunity to enter other State and National colleges and universities. SCAD staff will assist with short courses and lectures that inform students and faculty about smart agri-energy systems.

The smart ecocity will build an Ecolanda Eco-Learning Centre. The Eco-Centre will focus on sharing smart biosystem technologies with students, farmers and Ecolanda associates.

The Ecolanda Eco-Centre will listen and learn from indigenous farmers and share their methods with others. SCAD team members will help train indigenous farmers in sustainable practices that make their farms more productive.

The Ecolanda Eco-Centre will become an eco-tourism destination for people globally who are interested in biosolutions to save and repair our planet.

Education and training will be provided using in-person, distance learning and hybrid systems. Many distance learning offerings will be free in order to maximize reach and access.

Many ecocity residents work on the Ecolanda project. Associates and their families have access to excellent food, medical, transportation and communication services.

Ecolanda makes education and skill training available for all associates and their families, regardless of age or ethnicity.

## 4.  Abundant Agriculture

*Nature demonstrates abundance. Ecolanda celebrates and practices nature's richness.*

Ecolanda uses highly productive and resource efficient controlled environment agriculture, CEA. Controls allow the cultivation of substantial amounts of high-nutrient food, feed, bioenergy and other valuable bioproducts with zero waste and zero emissions.

SCAD agri-energy production employs thousands of people in good jobs. Ecolanda's commitment to diversity ensures that the project hires and trains women and men, young and old, able and handicapped as well as indigenous people. Associates enjoy access to excellent benefits and extensive training and development. Many associates have an opportunity for cross-training so that they can perform several roles.

SCAD uses some traditional agricultural methods but prefers sustainable abundant agri-solutions.

### Broadacre crops

Broadacre cropping uses minimal mined resources and instead substitutes biocycled nutrients in biofertilisers. Improved soil structure makes the soil more porous, which reduces the need for extensive cultivation by 50%.

The combination of enriched humus and subsurface drip irrigation reduces water consumption by 80%. Drip irrigation and Nrich biofertilisers nearly eliminate the need for inorganic mined fertiliser.

SCAD avoids pesticides and other poisons preferring instead to use SCAD's proprietary Nrich organic biofertiliser. Nrich biofertiliser contains plant hormones and other natural stimulants that allow the plant to produce its own natural biopesticides. The integration of biofertiliser and drip irrigation allows for zero fertiliser or water waste, which eliminates run off pollution to local waterways and well water.

The agricultural methods graphic below illustrates the continuum from Intensive Mechanical Agriculture, IMA, to Organic, Resilient and Abundance.

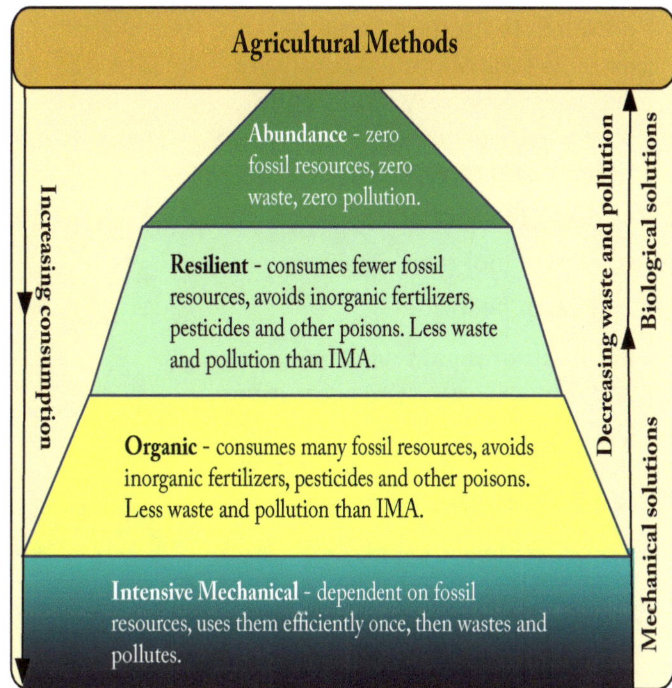


SCAD uses some IMA production, but continually innovates with new ways to use natural resources more efficiently. SCAD uses extensive organic production with smart innovations that reduce dependence on fossil natural resources. SCAD plans to help small-scale and indigenous farmers practice resilient and abundance methods.

SCAD's hydroponics, aeroponics, algae and other microcrop cultivation each practice abundant agriculture with the goal of using zero fossil resources, creating zero waste and zero pollution.

SCAD agri-energy operates carbon negative and water positive. Food cultivation uses nano and microcrops that are voracious consumers of $CO_2$. Each 100 tons of food captures 183 tons of $CO_2$.

Modern industrial crops cannot withstand saline soil or water. SCAD grows several energy and animal feed field crops that are productive with modest levels of saline in the irrigation water or soil.

### *Ecological costs*

Modern industrial agriculture uses mechanical methods powered by fossil fuels to force food production. Modern farming extracts tons of natural resources and uses them very inefficiently only once. Agri wastes are costly to farmers and to society:

- 50% of irrigation water
- 60% of inorganic fertilizer
- 95% of pesticides

Residuals erode on wind and water and foul air, water and ecosystems. Organic methods moderate some toxic waste, especially inorganic fertilizer and agri-poisons.

Ecolanda uses smart agriculture that saves, compared to industrial methods, 75% of water, 90% of inorganic fertilizer and 100% of pesticides.

Controlled environmental agriculture allows growers to consume no cropland and clean rather than pollute air and water.

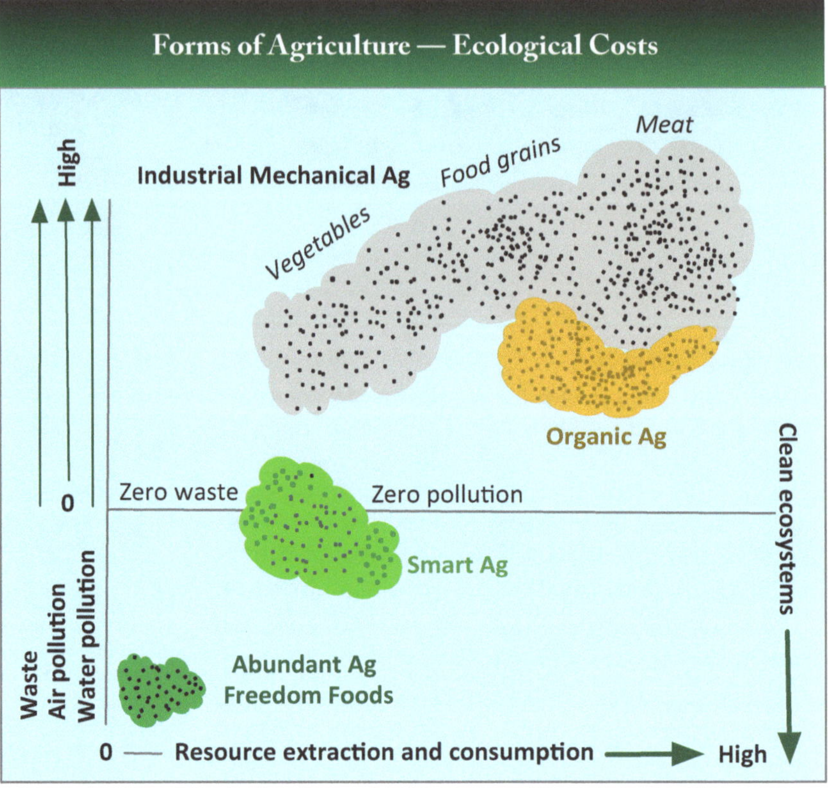


The Ecolanda architecture applies 7-generation biosystems that are living systems powered by free solar and other renewable energy. The circular linkages assure systems integration which eliminates waste and pollution.

While producing food many times faster than industrial agriculture and field crops like rice, maize and wheat, SCAD biosystem clean the air and water. One hectare (10,000 m$^2$) of algae microcrops can capture 6.7 tons/day of $CO_2$.

Ecolanda cultivates freedom foods that free growers from the substantial cost of many extracted natural resources and the waste associated with application.

Food grains like rice, maize and wheat deliver only 8 – 11% protein. A kilogram of rice consumes 1,000 liters of fresh water. Some algae species grow 20 to 50 times faster, consume zero fresh water and deliver clean edible biomass that contains 65% protein.

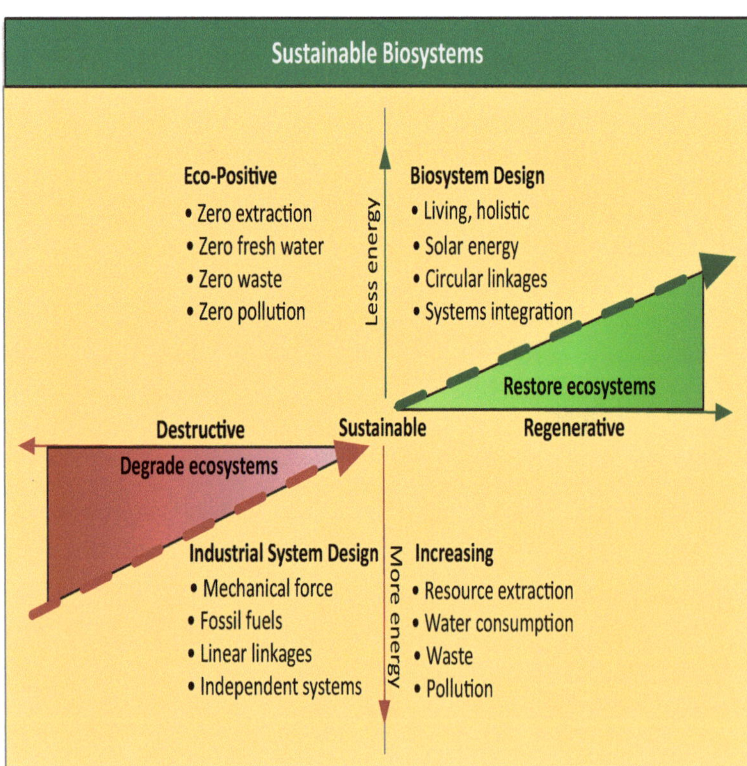

## Climate-controlled dairy

Beef and other meat animals grow with full waste capture and reuse. Animals are raised in large barns that capture gas, liquid and solid waste. Animal methane rises to the ceiling where it collects. A sensor switches on a pump to draw the methane to a combustion chamber with a turbine that generates electricity. Electricity is stored in an aluminium battery for later use on the farm.

An offtake from the combustion chamber flows the CO and $CO_2$ to an algae biosystem where the carbon is assimilated by microorganisms and transformed into new clean and nutritious feed.

Liquid wastes are collected and also flow to an algae biosystem. Those rich nutrients are transformed to valuable biomass that may be used for more animal feed or crop fertiliser. Manure goes through an anaerobic digestor. Off-gasses are biocycled in adjacent algae biosystems.

Biocycling animal waste streams eliminate external waste and enable zero emissions to air, water and ecosystems. Biocycling eliminates the need to extract fossil resources and reduces operational costs.

Phosphorus has become the limiting resource for food production. This finite resource has become increasingly scarce and expensive. Without phosphorous, no crops grow, and no children grow. Over 80% of mined phosphorus is wasted in the modern industrial food supply chain. Most the phosphorus waste flows into waterways creating dead zones.[10]

SCAD plans to capture and repurpose nutrients such as phosphorous up to 10 times, similar to Nature's path.

## Climate-controlled RAS aquaculture

Climate-controlled, land-based recirculating aquaculture systems provide unique advantages over ocean, estuary, lake or river farming. Land-based fish farms assure consistent, high specification water quality and temperature.

Climate-control allows for temperature and light variations that match the seasons, which is important for fish that naturally migrate.

SCAD filters biowastes from aquaculture systems and flows those wastes to hydroponic CEA where the nutrients are assimilated by vegetables.

Post hydroponic water still contains some waste nutrients and biosolids. This water flows to an algae biosystem for cleaning.

SCAD aquaculture uses state-of-the-art algae biofeed. Biofeed nutrition perfectly matches the fin or shellfish life stage. Sensors and monitors capture all the critical metrics, ensuring fish start healthy and stay healthy throughout the lives

SCAD creates specific recipes based on tested algorithms to optimize healthy aquaculture.

*Feed and observation aquaculture platform*

### Stressor management

Every animal grown faces stresses from a wide variety of sources. Most animal stressors are known and monitored. Field studies show that healthier nutrition significantly improves stress tolerance, which improves survivability and vitality.

Nrich enhances nutrition and ensures the animals receive the full set of macro and micronutrients, as well as vitamins, minerals and trace elements.

Algae-based biofeed are natural to fish, which improves both palatability and survivability. Algae are high in plant sugar, glucose, which makes the feed attractive. Superior nutrients are a waste if they were not bioavailable. Algae cells are so tiny, animals immediately assimilate the biofeed, which spurs growth at every life-stage.

High food bioavailability results in 10 to 20% less waste. Algae biofeeds are easier on animal stomachs than food grains. Efficient food uptake reduces producer costs because less food is needed. Less waste also lowers farmer waste management cost, which may be 20% of operational cost.

Many industrial producers dose their animals with antibiotics to reduce gastric distress from food grains and other foods unnatural to animal guts. SCAD has no need to use antibiotics because animals are able to quickly digest bioavailable biofeed.

### Aquaculture production

Ecolanda plans smart controlled environment RAS production of 100 million fingerlings annually. These include both fin and shellfish.

All the land-based RAS cultivation will operate with no emissions and no wastewater flows to external waterways. Most the water containing fish excretions flows to hydroponic CEA systems where plants will absorb the nutrients that went through the fish.

### RAS aquaculture process

The SCAD Nrich process flow follows the RAS aquaculture schematic:

1. Fish tanks      2. Filtration  3. Reservoir 1
4. Water lifting     5. Biological filtration
6. UV disinfection 7. Degassing  8. Reservoir 2
9. Oxygenation    10. Fish feeding
11. Water intake  12. PH & alkalinity
13. Temp control  14. Salinity control
15. Sludge treatment 16. Grading

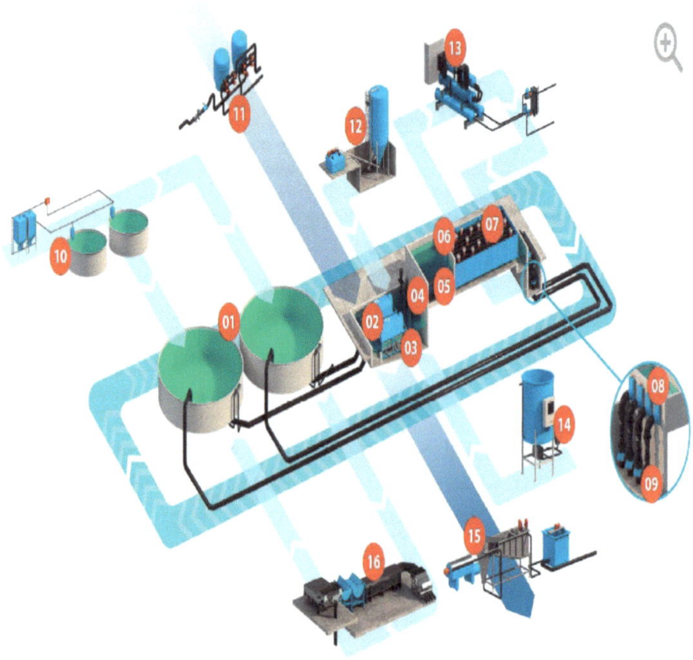

### Superior nutrition, superior animals

Nutritional and medical research reinforces the value of enhancing nutrition to improve health, stress-tolerance, disease avoidance, disease treatment and survivability. Microcrops offer superior nutritional attributes that are not available in land-based crops.

Succulents are plants that store excess water. No English word describes plants that store and deliver higher nutrient density, quality and diversity. **Nutralence** describes produce in a manner that offers relevant comparative metrics.

SCAD applies advanced metrics to monitor and assure high nutralence, which offers nutritional differences on five key dimensions.

**Nutralence metrics**

Nutralence includes these five elements:

1. **Nutrient quality**, the availability and stability of essential nutrients
2. **Nutrient density**, nutrients per bite
3. **Nutrient diversity**, the variety of nutrients
4. **Bioavailability**, the assimilation of nutrients easily and quickly without gastric distress
5. **Bioactive compounds**, over 300 bioactive compounds act to protect the consumer from disease or environmental stress

Nutralence metrics are shared with top nutritional institutions globally for validation. Studies have revealed significant positive nutralence differences compared with industrial agriculture crops.

| Nutralence Attribute | Algae compared to fossil foods or meat | Criticality |
| --- | --- | --- |
| Nutrient quality | 10x to 100x | No contaminates. No pesticide residuals. |
| Nutrient density | 10x to 100x | No empty calories. More nutrients per bite. |
| Nutrient diversity | 10x to 100x | More micronutrients, vitamins and trace elements. |
| Nutrient bioavailability | 10x to 100x | Faster and more reliable assimilation into the body. |
| Bioactive compounds | 100x to 1000x | More disease protection. More disease therapeutics. |

Each nutralence attribute typically measures 20% or higher than modern industrial food and feed crops in SCAD cultivation. The table illustrates the nutralence advantage from experience with multiple crops.

Nutrient quality may be most difficult to measure. In general, microcrops and field crops fertilized by algae biofertiliser deliver independent laboratory values that show the produce contains more than 10x more micronutrients and vitamins compared with industrial field crops.

Extensive research shows the value of higher nutrient quality, density and bioavailability.

## *Climate-controlled organic vegetables*

Vegetables, microgreens, fruit, nuts, seeds and herbs are grown in smart vertical farms. SCAD Ecolanda climate-controlled hothouse cultivation includes several highly productive non-soil technologies, hydroponics and aeroponics.

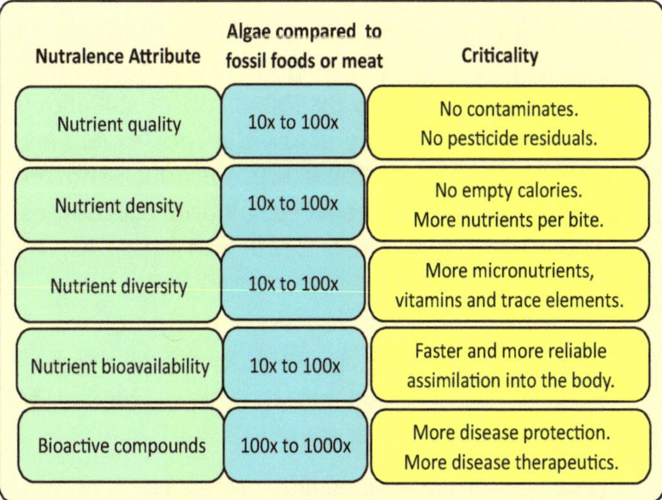

Controlled environments allow for rapid production of a wide variety of high-value crops. Cycle times for SCAD CEA cultivation exceeds field cultivation by 10 to 20 times.

### Vertical farms

Vertical farms reduce net ownership, building and operating costs over the 30-year expected life of each building.

Efficient vertical farms reduce blue water consumption 90% by minimizing evaporation and capturing transpiration. Land footprint drops by over 90% due to far higher productivity, vertical scaling and use of non-cropland. Ecolanda vertical farms reduce the total agri-energy ecological footprint by over 90%.

SCAD production of microgreens and barley shoots for animals use automated, climate-controlled hothouse cultivation to increase productivity and nutralence. Automated systems grow nutrient dense animal fodder efficiently.

SCAD applies advanced automation and robotic systems to maximize the reliability and efficiency in vertical farm cultivation. Automated systems move growing trays along at a prescribed rate. Robotic systems speed planting, harvesting and packaging.

*Vertical farm arrays and barley shoots*

### Broadacre drip

Broadacre drip irrigation systems improve water se efficiency by 75% because they avoid evaporation loss. Drip systems provide efficient delivery systems for nutritious algae biofertiliser.

SCAD broadacre drip irrigated crops include human food, animal feed and energy crops such as Arundo Donax.

Smart drip systems address disadvantages associated with desert and prairie farming.

Desert soil is often compacted and salty. SCAD grows some salt tolerant crops but even those crops produce better with less soil salt and looser soil that allow roots to extend deeper.

Algae biofertiliser has demonstrated the ability to improve soil porosity by 500%. Looser soil allows soil salts to percolate below the root zone. Field research with Del Monte Fresh Produce showed algae biofertiliser improved soil porosity 5 times, which and doubled root depth and strength.

### Processing facilities

Modern processing facilities typically produce massive waste products that must hauled to dumps and burned or buried.

Ecolanda processing facilities for all types of agri-production, including textiles, are designed for zero emissions and waste. Processing facilities include sorting and packing sustainably produced animal and plant products.

Processing facilities biocycle waste streams – gas, water and biosolids – into fresh feed, biofertiliser and other bioproducts. Waste products from processing a single cow or buffalo often exceed the edible portion of the animal. A futures trading team actively matches animal and plant biproducts with appropriate markets. Residual wastes are biocycled to recover the carbon and other nutrients.

The next section examines the Nrich process.

# 5.  Nrich – Superior Nutrition



Fossil fuel powers mechanical engines. Richer fuel improves engine performance.

Nutrition powers cellular metabolism; bioengines for plants, animals and people. Nrich bioenergy improves metabolism, health and vitality. Algae biomass and foods deliver superior nutritional energy for plants, animals and people.

Nrich biofertiliser enhances crop growth by >20%. Other attributes improve too such as quality, yields and taste.

Nrich biofeed enhances animal growth by >20%. Fish grow faster, excrete less waste and benefit from stronger stress tolerance.

Nrich creates plant-based meat with 2.5x the protein per bite as beef. The food is healthier for people and our planet. Animals are said to prefer non-meat alternatives too.

Algae, the first food on the planet offers healthier nutrition – lipids, proteins and carbohydrates. Lipids are long carbon chain molecules. Lipids store energy for the plant and serve as the structural components of cell.

Proteins are large organic compounds made of amino acids arranged in a linear chain connected by peptide bonds. The plant's genetic code determines the sequence of the amino acids, but nutrient limitations may cause changes to the production of amino acids. Most proteins are enzymes that catalyze biochemical reactions and plant metabolism. Other proteins maintain cell shape and provide signaling functions. Algae deliver more and healthier proteins per bite than any animal meat or any land plant.

Quality animal feed demands protein. Professor E.W. Becker wrote an excellent review article on algae animal feed. He concluded the quality of algae protein is equal, and in many cases superior to conventional plant proteins, including soy.[11]

Starches are complex carbohydrates which are insoluble in water. Plants use starches to store glucose as plant sugar.

Biomass composition among algae species varies tremendously. Some algae hold 80% lipids while others are 65% protein. Some macroalgae, seaweeds contain 92% carbohydrates. Species selection is critical, not just for the desired composition, but for a host of micronutrients and growth biostimulants that vary across cultivars.

Algae varieties offer almost infinite nutrition combinations and useful bioactive compounds. Special attributes such as omega-3 oil production can be enhanced through selection screens for naturally occurring organisms and mutagenesis, which is similar to a rapid hybridization process.

Some companies hybridize algae to express more, or less of target compounds. Algae biofeed will become far more desirable as specialty compounds that help animals with digestion, biosorption and protection against pathogens are discovered and used in algae cultivars.

Algae biofeed provides the full set of essential nutrients for animals. Considerable research focuses on protein quality, which varies across various species of terrestrial plants and algae. Unsurprisingly, different types of animals grow faster and better with high quality proteins combined with a full set of micronutrients.

Algae composition displays considerable variation with seasons. Algae biofeed formulation requires continuous monitoring, similar to field grains. Traditional crops such as maize are often mixed with antibiotics to improve digestion. Other supplements such as omega-3 fatty acids may be added to improve animal health.

Algae biofeed can eliminate pharmaceuticals as feed additives because digestion does not cause problems. Various algae species provide different amounts of omegas and other specialty oils.

No terrestrial crop or animal produces omega-3 fatty acids naturally. Some have been genetically modified to produce a low-quality omega antioxidant, but research has not established whether the molecule is bioavailable to animals.

Algae biofeed can deliver the omegas of choice.

1. **Linoleic acid, LA,** is a non-essential unsaturated omega-6 in soaps, emulsifiers, quick-drying oils and a variety of beauty aids.

2. **Arachidonic acid, AA**, is an omega-6 fatty acid that is not essential. It may moderate or induce inflammation and plays a role in the operation of the central nervous system.

3. **Eicosapentaenoic acid, EPA,** an omega-3 fatty acid that gives the same benefits as fish oil.

4. **Docosahexaenoic acid, DHA,** an omega-3 fatty acid is the most abundant fatty acid found in the brain and retina. DHA deficiencies cause cognitive decline and higher brain cell death.

Many algae species are tolerant of wide variations in growing conditions. Some species are nearly blind to geography. All algae species can be grown in CEA vertical farms.

### *Algae biofeed nutrition*

The nutritional contents of algae are superior to land-based animal feed. Algae biofeeds provide a renewable substitute for conventional animal feed ingredients for meat, dairy, poultry and fish.

| SCAD Biofeed Animal Nutrition and Productivity | |
|---|---|
| **Animal nutrition:** | **Animal productivity:** |
| • Stronger vitality | • Faster growth, 10% |
| • Higher nutralence, 30% | • Higher survivability, 10% |
| • Higher palatability, 20% | • Stronger vitality, 20% |
| • Bioavailable nutrients, 40% | • Higher stress tolerance, 20% |
| • Less waste, 10% | • More robust health, 20% |

Algae contain all the vitamins, minerals and micronutrients needed for healthy biofeeds, including Vitamins A, B1, B2, B6, B complex, C, D, and E. Algae are rich in niacin, iodine, potassium, iron, magnesium and calcium.

| SCAD Algae Biofeed Compared with Field Grains | |
|---|---|
| **Higher productivity:** | **Ecological savings:** |
| • Growth rates, 20 to 100% | • Ecological efficiency, 80% |
| • Productivity / ha, 50 to 100% | • Less waste, 80% |
| • Protein / ha, 50 to 100% | • Water efficiency, 50 to 80% |
| • Omega-3s, 100% | • Diesel fuel efficiency, 90% |
| • Quality, 20% | • Inorganic fertilizer, 80% |

Algae biofeed grows with substantially higher productivity per hectare than grains or palm oil. Each algal cell contains polysaccharides, sugar and starch, as well as iron, sodium, phosphorus, magnesium, copper and calcium. Algae biofeeds deliver nutritional benefits not available in land plants, antioxidants and bioactive compounds.

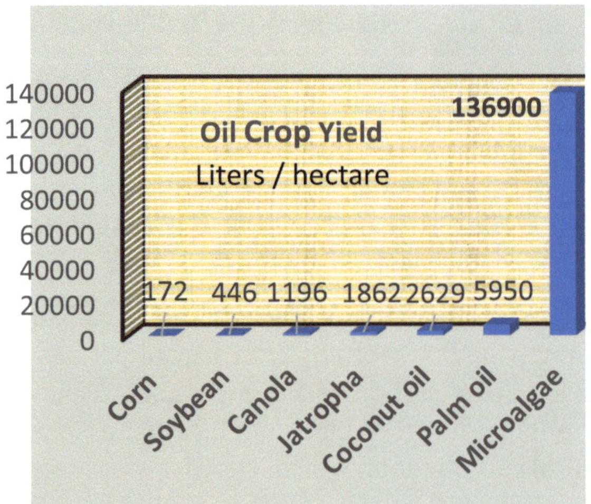


Many valuable biomolecules in algae are not synthesized in the animal, (or human) body. However, they are considered essential for healthy body, brain and heart growth and development. Therefore, animals must get them through their diet.

Algae biofeeds provide a rich source of high-quality protein, vitamins, micronutrients, trace elements and carotenoids.

Algae produce valuable biomolecules including astaxanthin, lutein, beta carotene, chlorophyll, beta-1,3-glucan and phycobiliprotein.[12]

Algae provide both an ideal nutrient package and delivery system. The algae package, a cell, is so small it becomes immediately bioavailable to the animal. The algae delivery system provides higher nutralence than conventional animal feed, with significantly higher nutrient quality and density,

## Nrich biofertiliser

Algae biofertiliser significantly improves seed germination, early growth, speed of growth to maturity and survivability by more than 20%.

Multiple projects globally have demonstrated SCAD biofertiliser improves produce yields in excess of 20%. In addition, produce display superior nutrition and avoid hidden hunger and micronutrient deficiencies.

Improvements in produce market value arise from the unique Nrich nutrient delivery system.

Nrich biofertiliser delivers the full set of 43 macro and micronutrients plants need for successful growth. Nrich also delivers key vitamins, minerals and trace elements, which improve crop stress tolerance as well as colour, taste and texture.

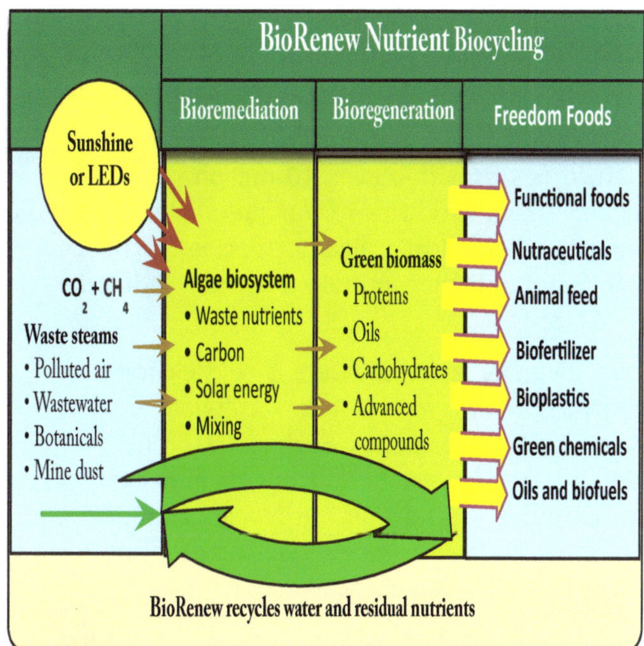


Nrich relies on BioRenew for nutrient capture, recycle and repurpose.

Larger produce with a longer shelf life demonstrates that the full nutrient set delivered by algae biofertiliser improves produce quality and health. Only healthy produce can resist spoilage for several extra days. Farmers often receive higher market prices for early produce.

| SCAD Biofertiliser Compared with Industrial Fertiliser | |
| --- | --- |
| **Improves crop:** | **Improves market value:** |
| • Germination rate, 20% | • Taste and aroma, 20% |
| • Time to maturity, 20% | • Vitamins & minerals, 50-100% |
| • Health and vitality, 30% | • Digestible nutrients, 50-100% |
| • Yield and quality, 30% | • Color and texture, 20% |
| • Produce size, 20% | • Shelf-life, 25% |

Nutritional and productivity increases improve grower yields and profitability. Ecolanda provides a plentiful continuous cycle of new waste stream nutrients. Nutrients may be captured from air, water and biosolids including animal manure or coal dust.

Most modern farms constantly extract nutrients but replace only three macronutrients, nitrogen, phosphorus and potassium.

These nutrient waste streams serve as building blocks for new green biomass that may be transformed into biofertiliser, feed or a wide array of valuable bioproducts.

| SCAD Biofertiliser Compared with Industrial Fertiliser | |
| --- | --- |
| **Reduces production costs:** | **Enhances soil:** |
| • Tillage, 30% | • Porosity, looseness, 500% |
| • Diesel fuel, 40% | • Microbial life, 500% |
| • Irrigation water, 25% | • Erosion resistance, 30% |
| • Inorganic fertilizer, 80% | • Bioavailable nutrients, 50% |
| • Pesticide / herbicide, 90% | • Organic material, 20% / year |

Growers may grow the same crop for months or change the culture every month or every season to a different algae species that provides more of the desired target compounds.

## Microcrop diversity

Scientists estimate there are about 300,000 land-based plants. Only 17 low-nutrient plant species are consumed as 90% of the global human diet.[13]

Nature has provided over 10 million species of algae.[14] Each produces its unique set of target compounds. Most algae species produce significantly higher nutralence than industrial field crops.

This diversity gives growers a wide choice of cultivars. Most growers today cultivate one or more of 30 algae species. Each cultivar may offer higher or lower productivity and target compounds based on growing conditions. Improved cultivation models will increase species diversity and productivity.

Ecolanda growers avoid the use of pesticides and poisons. Algae biofertiliser provides a wide array of plant hormones and biopesticides that protect the plants. Avoidance of pesticides assure that no poisons enter the local waterways or tag along as residuals on consumer produce.

## Algae use solar energy

Algae make up the first step on the food chain. Each algae cell contains all the crucial elements needed for healthy growth, development and reproduction.

The mechanism for action, photosynthesis, transforms water and carbon dioxide to plant sugars in a carbon-rich biomass. The only emission is pure oxygen released in in enormous quantities. The oxygen can be captured and reused in other processes.

Photosynthesis uses free sunshine for energy. Energetic photons streaming with sunshine drive the process. Algae and other plants are agnostic regarding the photon source. Growers may supplement sunshine from other indirect solar energy sources such as fibre optics, lasers, mirrors or LED lights.

Alternative photon sources provide a critical cultivation bridge in areas that do not have the benefit of many clear sunny days. Vertical farms use LEDs and other non-direct solar sources to deliver photons to the crops.

## Single-celled biofactory

Algae provide nature with the fastest growing plant on the planet. Algae do not have to waste energy on roots, leaves and stems.

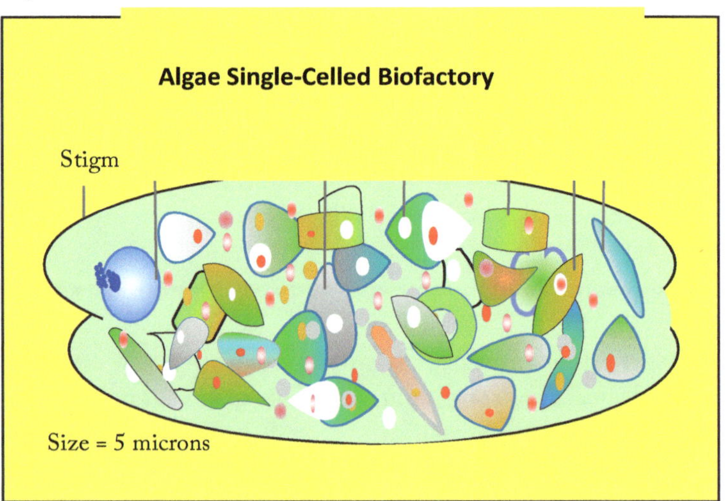


The single-celled algae biofactory serves as the Ecolanda workhorse. The single algae cell appears quite simple. However, billions of years of evolution have enabled the tiny plant to create a simple but effective array of biomechanisms. These features act as an integrated biofactory for the production of thousands of bioproducts.

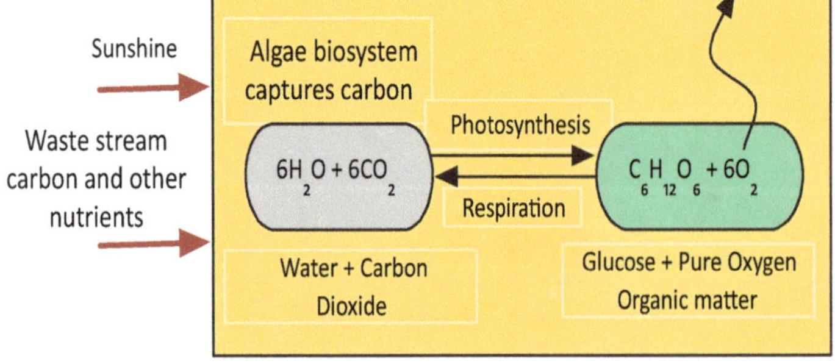

Terrestrial plants uptake nutrients through their roots that flow through their stems and leaves. Rooted plants cannot use wastewater due to likelihood of salt ions that would clog their circulatory system and kill the plant. Algae have no roots so salt ions are not a problem.

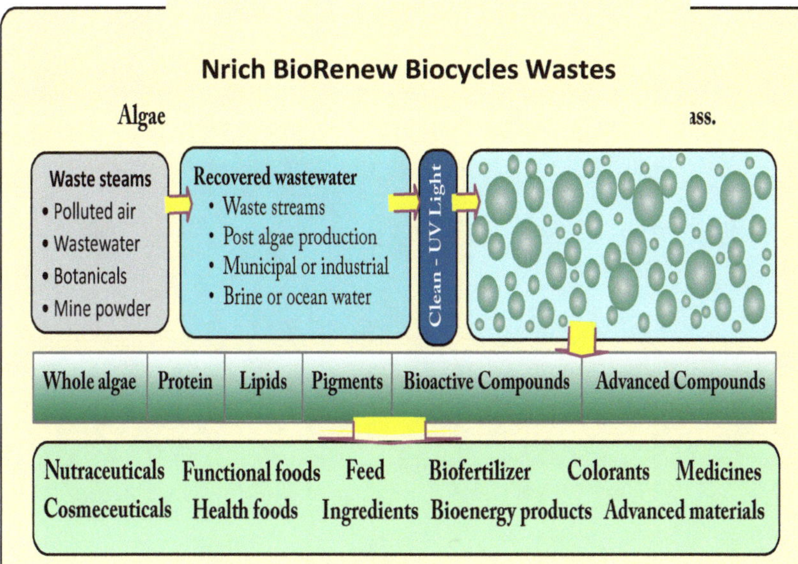


Recovering nutrients with land plants would be a loser's game. Terrestrial crops waste over half the applied nutrients, which makes the concept infeasible. Algae use nutrients in the culture very efficiently. The soft green biomass makes it relatively easy for growers to extract compounds.

## Dormancy

Terrestrial plants are extremely limited because they are constrained by their roots. Land crops had to adapt and grow even if the soil lacked certain nutrients. This "hidden hunger" effect means that field crops may look big and bright but lack nutrients that were unavailable in the soil.

When terrestrial plants evolved from algae 500 million years ago, they made many sacrifices and left behind extensive capabilities, including the ability to go dormant. If a field crops lack inputs, the plants do not have a choice between dormancy and death. Field crops lacking water or extreme heat or cold, die and die quickly.

Algae do not exhibit hidden hunger because they can go dormant for decades. Each algae cell assimilates the full set of essential nutrients. When the first nutrient is unavailable in the culture, algae growth stops.

The cells remaining in the culture are loaded with nutrients, but growth and propagation pause. Each cell carries the full nutrient set, which is why algae nutralence significantly benefits consumers whether eaters are people, animals or plants.

Growers may stress an algae culture by withholding nutrients or changing culture parameters. Algae biofactories respond by making new, often valuable compounds such as astaxanthin.

## Growth Speed

How do algae grow so fast? Energy distribution provides the answer. Field grains are multicellular organisms that invest most their energy in non-food components necessary for survival. Over 90% of biomass growth – roots, circulation system, structure and even sexual apparatus – consumes lots of energy but grows no food.

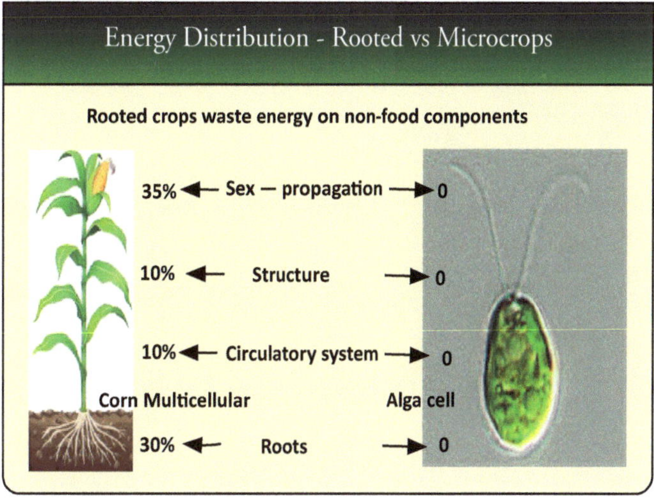


Algae are single-celled, water-based plants without roots or circulatory systems. Not only do algae plants grow 50 times faster than field grains, but the biomass also contains over 90% food and other valuable compounds.

# 6. Renewable Energy Generation

*Renewable energy creates jobs,
reduces consumer energy costs
and cleans environments.*

Renewable energy drives critical actions in each Ecolanda vertical segment, agri-energy, waste, water and ecocity.

SCAD draws from a suite of renewable energy technologies to ensure continuous and reliable power. Ecolanda consumes considerable non-fossil energy across the five sectors.

Geography, physical eco-features and climate factors determine the selection of specific renewable energy technologies.

## Solar energy

Solar energy includes photo voltaic, con-centrated solar and others. Photovoltaic systems generate electricity directly from sunlight via semiconductors.

Electrons are freed by solar energy and travel through an electrical circuit.

Concentrated solar power, CSP, generates solar energy by using mirrors or lenses to concentrate a large area of sunlight onto a receiver.

CSP energy can be stored and delivered night or day, whenever needed. Solar power systems can be sited on desert or non-cropland. They require minimal water or other natural resources.

## Energy flows

Several wind energy systems are options. Selection depends on the local microclimate, prevailing winds, fetch and diurnal wind speeds. Horizontal axis turbines rotate with the

force of the wind and transform the airstream into electrical energy.

Vertical wind systems may offer more efficient energy production with lower wind speeds. Vertical wind systems may be conveniently sited locally on buildings and in the ecocity.

All green energy systems must be designed to survive extreme weather events including heat, typhoons or hurricanes.

Hydro energy systems include river currents, waves and tides, depending on river and ocean geology. The turbine on the right may be used for river flows or ocean tides.

A wave and tidal hybrid design captures energy from bouncing up and down and capturing lateral incoming and outgoing tidal energy.

Tide energy may be productive but is only possible with a flat coastal shelf and high tidal swing.

The following graphic illustrates SCAD's options for zero emissions sustainable energy.

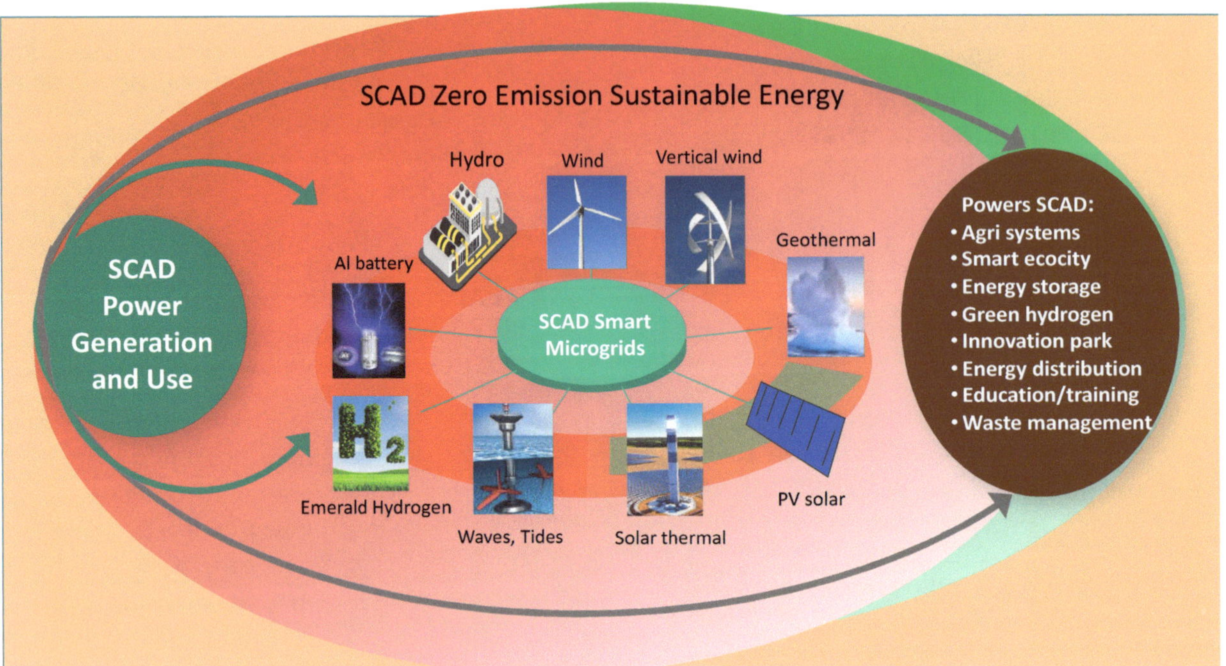



## Geothermal

Volcanic geological formations may allow geothermal if hot magma is accessible. Like river currents, waves and tides, geothermal produces energy consistently 24/7. Geothermal energy production systems may be sited distant from the Ecolanda location because geology and magma access dictates sites.

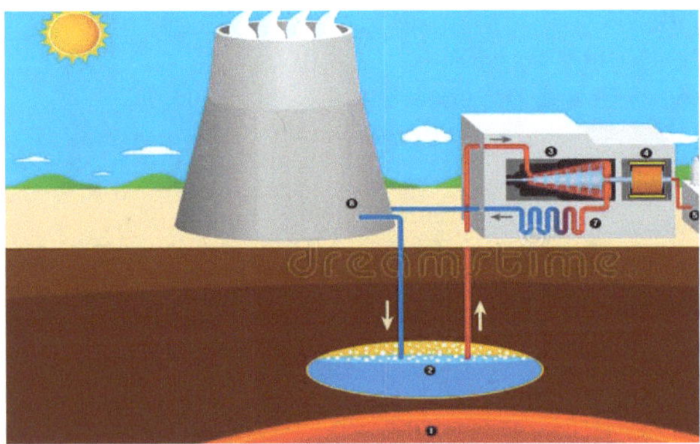

## Anaerobic digestion

Anaerobic digestion can transform organic waste, which is prolific in agriculture, to generate significant amounts of local energy. Digestion engages microorganisms to break down organic matter such as manure without oxygen.

Microorganisms assimilate the active solids and excrete biogas, mostly methane and carbon dioxide. The non-methane components are separated so the methane can be combusted as an energy source.

Certain types of organic matter break down more easily than others. Highly digestible organic matter generates more biogas faster. Fats, oils, and grease break down quickly while organic biosolids and sludge process more slowly.

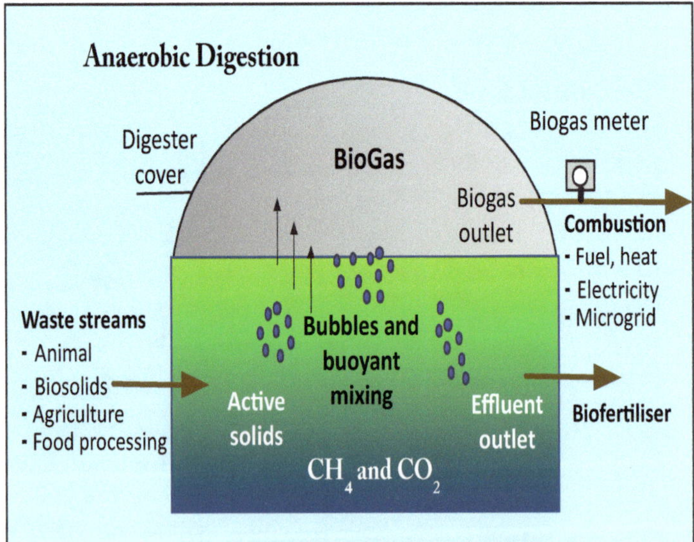

Codigestion can break down multiple types of organic waste in one digester. Codigestion might use manure with residential or restaurant food waste, food processing waste or byproducts. It may use oil and grease from restaurants or energy crops and crop residues. Codigestion often increases biogas production from difficult-to-digest organic waste.

Biogas may flow directly into a turbine generator for electricity or heat. Biogas may be converted into compressed natural gas, CNG and used as a transportation fuel. Digested solids may be used in fertiliser, landscape products, compost soil amendments or biodegradable planting pots.

## Biomass-to-energy

Biomass contains stored energy derived from the sun. Plants absorb solar energy and convert $CO_2$ and water into energetic biomass. The stored energy from the plant biomass below can be transformed into usable energy.

Biomass burned directly may heat homes, buildings, or water. Biomass combustion can drive a steam turbine. Biomass may be transformed to green hydrogen or green ammonia with zero emissions.

Microcrops are typically not burned to create energy because combustion volatizes the nutrients. Microcrop biomass offers substantially higher value in derivative bioproducts than energy products. Food is the most valuable stored energy on the planet. Food usually offers far more value than energy products.

## Energy crops

Ecolanda will cultivate several energy crops like Arundo Donax. This tall grass looks like a reed and grows well in temperate and subtropical regions. SCAD selected Arundo Donax as an energy crop due to its capacity to grow vigorously in marginal land that is too degraded or contains too much salt to grow other crops.

Energy crops undergo pyrolysis where they are heated in the absence of oxygen. The biomass does not combust because no oxygen is present. The chemical compounds, lignin, cellulose, and hemicellulose thermally decompose into combustible gases and charcoal. Combustible gases can be condensed into a combustible liquid called pyrolysis oil, bio-oil.

Energy crop pyrolysis produces energetic fuels such as charcoal, bio-oil, renewable diesel, methane, and green hydrogen.

Pyrolysis charcoal, often called biochar, works effectively as a slow release fertiliser. Biochar reduces chemical fertiliser requirements and can provide soil nutrients for 20 years or more.

## Biomass gasification

Biomass gasification uses a controlled process involving heat, steam, and oxygen to convert biomass to green hydrogen, ammonia and other products, without combustion.

SCAD uses Life Cycle Analysis, LCA, extensively to verify carbon capture and utilization. Growing biomass removes substantial $CO_2$ from the atmosphere.

The heat and steam release zero carbon because SCAD uses renewable energy to power gasification. LCA shows the net carbon emissions can be zero with SCAD's CCU and storage.

## *Waste-to-energy*

Ecolanda agri-production creates massive amounts of waste that flow in one of two directions:

1. Biological wastes go through the Nrich BioRenew process to recover, recycle and repurpose the valuable nutrients.
2. Some biowaste and other trash is transformed to energy products such as green hydrogen or green chemicals.

Botanical and animal waste provide the nutrients that are biocycled into fresh animal feed and biofertiliser.

Industrial farmers cannot put the animal waste directly on crops because far too many pharmaceuticals from the animals are likely to find their way into crops. Ecolanda growers do not have this problem because animals receive few or no medications. In addition, algae cells are too small to assimilate a large pharmaceutical molecule.

Algae biocycle wastes and break down the large molecules by stripping away individual elements. Algae may biocycle hazardous wastes in a manner that removes the toxins and other hazards, including radioactive compounds.

Special biosystems are designed to remove poisonous heavy metals such as lead, cadmium, mercury and arsenic from wastewater before another biosystem cleans the water of organics.

Biological processes are better positioned to biocycle nutrients from waste gas, water and botanical wastes while heat and pressure serve effectively for MSW and industrial waste.

Non-biological wastes are transformed to syngas that can make green hydrogen, green chemicals or other energy products for energy use locally or for export.

SCAD's hybrid waste-to-energy uses bioactive microbes to break down material into syngas or green hydrogen, ammonia or chemicals.

SCAD brings all types of unsorted wastes by rail from a few hundred kilometers around Ecolanda. The MSW and industrial waste comes from cities, towns and rural communities. SCAD targets 10 trains daily with 40 x 100 metric ton compressed wagons.

SCAD employs a technology for waste-to-syngas (energy) that does not require waste sorting. Heat and pressure can transform waste to a variety of energetic products, below.

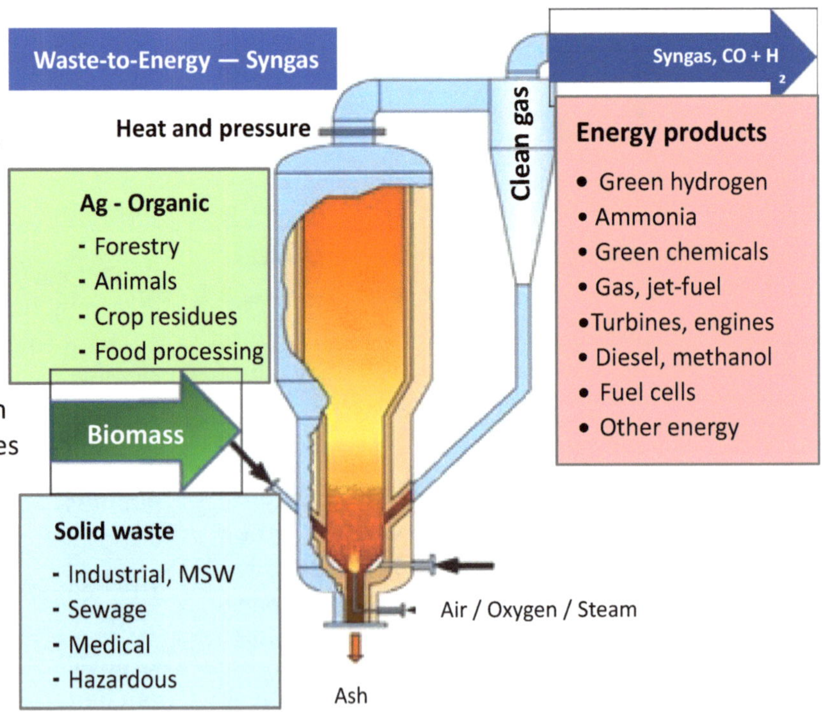

### Waste-to-syngas

SCAD scales waste-to-energy centres to fit waste availability. Each unit design aligns with the commitment to manage waste with zero emissions and zero pollution. SCAD can adjust the process to make a wide array of specialty green chemicals or energy products that current or futures markets value highly.

The process can handle virtually all types of waste and waste sources. Minimal waste sorting is required. The gasification process occurs very efficiently in a single chamber with no moving parts. Therefore, the equipment can work 24/7 with minimal maintenance.

SCAD waste-to-energy systems capture and reuse all emissions. Pyrolysis and combustion off-gasses are captured and flow to algae biosystems that biocycle the carbon and residual nutrients.

Naturally, the ash or any other residuals are constantly tested for toxins. Residual ash, typically less than 10% of the total waste, may be sequestered into green concrete or sustainable building materials such as resin composites.

Syngas, or synthesis gas, is a fuel gas mixture containing hydrogen, CO and typically $CO_2$. Syngas often serves as an intermediate in creating synthetic natural gas, SNG.

Syngas can be formulated into aviation petrol, fuel for energy production or diesel engines. Syngas can create sustainable green chemicals for industry or green hydrogen for transportation.

Green chemicals are designed with biological processes that reduce or eliminate the use or generation of hazardous substances. Green chemicals reduce the negative impacts of industrial chemicals on human and animal health and the environment.

### Bio-based chemicals

Green chemicals promise fewer emissions and biodegradability. Current green chemicals are used to make biodegradable plastics, superabsorbent sanitary products, bio-nylon, perfumes, skin creams, and detergents. These products create billions of dollars in revenue.

Most bio-based chemicals are currently made from petroleum derivatives combined with some plant material. Bio-based chemicals include:

- **Lactic acid:** Used to make PLA, which can be used for "biodegradable" plastics, which only biodegrade under extreme conditions.
- **Glucaric acid:** Prevents deposits of limescale and dirt on fabric or dishes, providing a green replacement for phosphate-based detergents
- **Muconic acid:** It's derivatives could replace non-sustainable chemicals used in the production of plastics and nylon fibres.
- **2,5-Furandicarboxylic acid (FDCA):** A stronger alternative to PET, which is used to make plastic bottles, food packaging and carpets.
- **Levoglucosenone:** A safer alternative to toxic solvents used in pharmaceutical manufacturing, flavours and fragrances.
- **5 Hydroxymethyl furfural (HMF):** A building block for plastics and polyesters
- **Itaconic acid:** A replacement for petroleum-based acrylic acid used in high-performance marine and automotive components.
- **1,3-Butanediol:** A building block for high value products including pheromones, fragrances, insecticides, antibiotics and synthetic rubber.
- **Levulinic acid:** Used in the production of environmentally friendly herbicides, flavour and fragrance ingredients, skin creams and degreasers.
- **n-Butanol:** Used in a wide range of polymers and plastics, as a solvent in chemical and textile processes and as a paint thinner

Bio-based chemicals represent a dynamic area of innovation and entrepreneurial opportunity.

Ecolanda entrepreneurs will make bio-based chemicals from waste-to-energy syngas and from algae oil.

## Bio-based concerns

All green chemicals are not produced equally eco-friendly. Many are worse for human ecosystem health than petroleum-based chemicals. PLA release harmful nanoparticles into the air during production and consumer use that cause serious respiratory problems.

Too many green chemicals, such as PLA, are not biodegradable except under extreme conditions, temperatures above 58 °C (136 °F). Few cities have the infrastructure to deal with bioplastics, so they often end up in landfills. Deprived of oxygen, bioplastics release methane, a greenhouse gas 23 times more potent than $CO_2$. When PLA becomes littered or dumped, biodegradation may take 60 years. When PLA floats into seawater, it does not seem to biodegrade at all.

A recent life cycle analysis, LCA study compared seven traditional petroleum-based plastics with four different bioplastics, below.

Scientists concluded that bioplastics resulted in significantly more pollutants.[15] Pollution comes from field crop fertilizers and pesticides and fossil fuels. Additional pollution occurs with the chemical processing needed to turn organic material into plastic. Bioplastics also contribute more to ozone depletion than petrol-plastics.

Bioplastics feedstocks require extensive land use which competes with food crops. Bioplastic proponents argue that the field maize, corn, is industrial and not human edible. While that is true, the crop still requires extensive cropland and blue water for irrigation. Several bioplastics impose contaminates on ecosystems such as carcinogens and other toxic "forever chemicals" pose human and animal health risks for decades.

Bioplastics often score worse on LCA than petrol-based chemicals because they combine the negative environmental and health impacts from agriculture, fossil fuels and chemical processing.

## Emerald biochemicals

Ecolanda creates healthy emerald biochemicals. These biochemicals are designed to benefit the health of people, producers, biodiversity and our planet. Emerald chemicals go beyond net-zero emissions and toxins as they also work to clean and restore health to our environment.

Emerald biochemicals ensure that the compounds and materials are made in a manner that is:

1. Completely free of fossil resources such as cropland, blue water, fossil fuels, inorganic fertilizers, pesticides or other toxic compounds.
2. Eco-friendly with zero waste, zero emissions and zero pollution.
3. Easily biodegradable with reusable carbon and other nutrients at the end-of-use.
4. Continuously cleaning air and water, biocycling carbon and restoring environmental health.

Emerald biochemicals undergo extensive LCA using both environmental and health metrics. They are tested to ensure safety for human and animal health, such as cancer and reproduction.

A specialty laboratory assesses the degree to which emerald chemicals support environmental health in their production, use and disposal. These tests look for dangers from physical hazards, such as flammability and reactivity and to toxicity for people, animals and biodiversity. Environmental reports also track possible compound persistence and bioaccumulation.

Ecolanda makes environmental reports available to regulators and to scientists for research purposes. Emerald chemical eco-metrics are important to share how clean biochemicals can support healthy society, economics and degraded ecosystems.

The next section examines Ecolanda's novel renewable energy storage solutions.

# 7. Renewable Energy Storage

*Imagine using sunshine for all our energy needs. All this requires is:*

1. *Collecting sunshine*
2. *Storing sunshine*
3. *Delivering sunshine*
4. *Using sunshine safely.*

*Hydrogen makes this possible.*

Renewable energy offers many advantages but energy flows are not constant. How can fluctuating energy sources such as sun and wind power existing grids, agri-industrial processes, manufacturing, heating and cooling and flexible transportation?

Energy storage technologies allow renewables to power the electrical grid during peak demand. Other times renewables build energy stores that can be used at any time for nearly any purpose.

Energy storage needs to be clean, require minimal additional energy and be safe in storage and use. Storage systems should integrate as seamlessly as possible with existing energy sources.

Ecolanda employs two strategies that safely and efficiently store variable energy from solar energy flows:

1. Transform electrical energy into liquid or gaseous chemical energy. Use stable and high-density forms such as syngas, green hydrogen and green ammonia.

2. Store energy directly in advanced aluminium-ion, Al-ion batteries.

Ecolanda renewable and waste-to-energy systems create clean energy that can be used immediately or stored for adaptive uses.

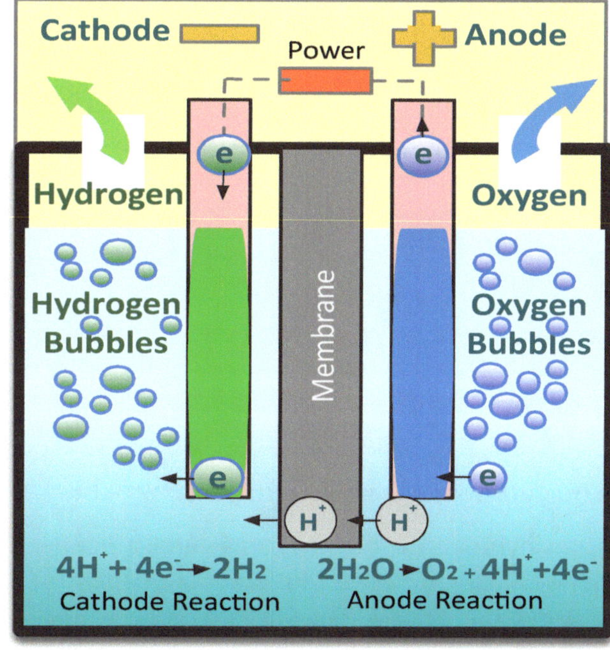


*Hydrogen*

Nearly all the energy on earth comes from solar energy directly through photons, pressure differentials with wind, or indirectly with when photons are collected and stored in biomass.

Hydrogen, H, offers an attractive natural energy storage package. It contains more energy per unit mass than fossil fuels. Hydrogen, the most abundant chemical in the universe, $\approx$ 75% of all baryonic mass. It is the simplest element on the Periodic Table with an atomic number of 1 and mass of 1.

Hydrogen is not a technology but a good energy carrier. It contains no carbon. Hydrogen burns with only one emission – water. Clean burning makes $H_2$ a clean fuel.

Hydrogen production uses simple electrolysis in water with an electrolyser. Emissions are pure oxygen and pure hydrogen gas.

## Colours of Hydrogen

Hydrogen production can be clean, dirty or very dirty, depending on the power source. A color spectrum defines various approaches.

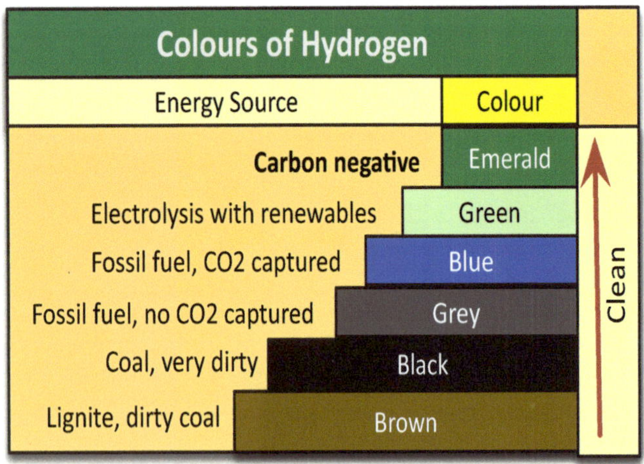


LCA shows that hydrogen production with fossil fuels largely defeats the green purpose. A clean-burning fuel that generates huge carbon emissions early in its lifecycle due to the use of fossil fuels for electrolysis creates a large carbon and eco-footprint.

About 70 million tonnes of mostly grey hydrogen are produced globally each year with natural gas. Green hydrogen could reduce these $CO_2$ emissions by 830 million tons. This equals the $CO_2$ emissions of the United Kingdom and Indonesia combined.

## Green hydrogen

Green or renewable hydrogen — made from the electrolysis of water powered renewable energy — serves as a key to climate neutrality.

The only carbon emissions are from those embodied in the generation infrastructure such as manufacture of wind turbines or solar panels.

Green hydrogen can supply low-carbon energy over very long distances. Renewable energy can be stored in green hydrogen to meet daily, weekly or monthly imbalances in supply and demand.

Green hydrogen can integrate the largely separate energy systems that are used today for heat, power, agri-industry and transportation.

Green hydrogen does three things:

1. Store surplus renewable power when the grid does not need the energy.
2. Decarbonize hard-to-electrify sectors such as long-distance transport and heavy industry.
3. Replace fossil fuels as a zero-carbon feedstock in chemicals, manufacturing and liquid fuel production.

Green hydrogen use has been blocked by supply and value chains that currently are too complex and uncoordinated. Retrofitting existing infrastructure generally requires time and costs that halt clean energy and clean fuel expansion.

Ecolanda does not suffer the challenges of remaking the human-built environment. Ecolanda will design and build simplicity and integration into the infrastructure for the four keys to sustaining hydrogen's green label: supply, storage, distribution and use.

Ecolanda also will produce and use ammonia, another way to safely store sunshine.

## Green ammonia

Ammonia, $NH_3$, is a colorless gas with a distinct odor composed of nitrogen and hydrogen atoms. It is produced naturally in the human body and in nature—in water, soil and air, even in tiny bacteria molecules.

Ammonia has such a high energy content, the weapons industry used $H_2$ extensively during

WWII as a prime component in high explosives like TNT. After the war, the weapons industry transformed into the ammonia fertilizer industry. Ammonia was the worst kept secret to the Green Revolution in agriculture.

Over 80% of the ammonia produced today serves as the powerful nitrogen fertilizer in agriculture. Unfortunately, as much as 50% of the ammonia applied to fields volatizes and creates toxic nitric oxides that form smog. Ammonia finds use as a refrigerant gas, in water purification and in the manufacture of plastics, explosives, textiles, pesticides, dyes and many other chemicals.

Green ammonia offers options in the transition to net-zero carbon dioxide emissions. These include:

**Energy storage** – ammonia stores easily in bulk as a liquid at low pressures, (10-15 bar) or refrigerated to -33°C. This offers a model chemical store for renewables. An extensive storage and distribution network already exists. Ammonia may be stored in refrigerated tanks and transported around the world by pipes, rail and road tankers and ships.

**Zero-carbon fuel** – ammonia can be burned in an engine or used in a fuel cell to produce electricity. Ammonia's only by-products are water and nitrogen. Maritime large ships, large trucks and rail industries are likely early adopters, replacing high-carbon fuel oil.

**Hydrogen carrier** – many applications may plan to use hydrogen but find the infrastructure for storage and transport infeasible. Ammonia stores and transports substantially easier and cheaper than hydrogen. Ammonia can be quickly and easily cracked to supply hydrogen.

Few gas, liquid or solid energy options are available with the promise of net-zero $CO_2$ emissions. Net-zero emission may be accomplished by truly creating no $CO_2$ emissions.

More commonly energy supply chains emit carbon at one stage in their lifecycle and other processes off-set the emissions with other CCU processes that may even be separate from energy supply chain. Accurate accounting for $CO_2$ emissions remains problematic.

SCAD has developed a proprietary carbon negative green hydrogen brand, **Emerald H₂**. Carbon negative means that every unit of Emerald $H_2$ energy captures and reuses carbon in lieu of allowing $CO_2$ emissions.

### *Emerald H₂*

SCAD's sets a high bar for carbon and eco-footprint. Other $H_2$ producers make green claims but full LCA shows $H_2$ production scores neither as clean nor as eco-friendly.

SCAD's extensive systems integration allows Emerald $H_2$ and Emerald Ammonia production at lower cost while demonstrating stronger carbon and eco-footprints. Emerald $H_2$ allows zero emissions through its life cycle including feedstock, production, storage and distribution.

Emerald $H_2$ production uses non-potable, saline water and an organic electrolyte, which adds value to its eco-footprint. SCAD captures and flows the substantial released oxygen to aquaculture to oxygenate fishponds.

Emerald $H_2$ provides numerous uses for stored solar energy in energetic gas or liquid form. Emerald $H_2$ may flow to power a turbine to create electricity for the grid or local industry.

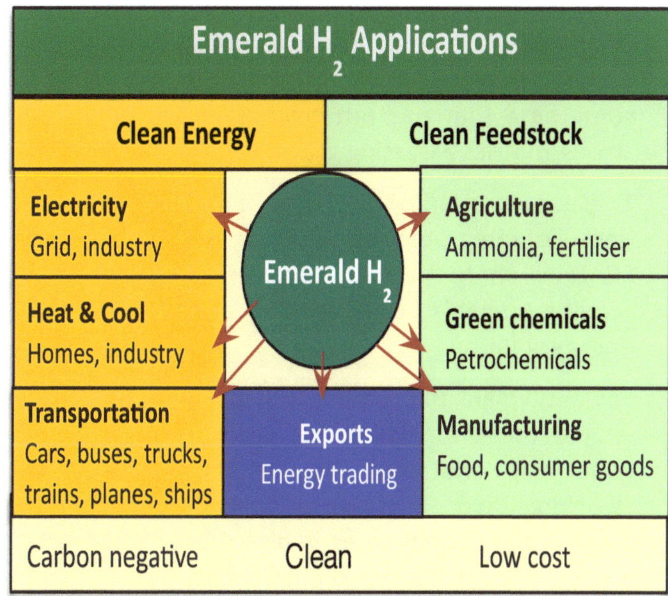

Hydrogen gas may be used for heating and cooling. Emerald H2 as a liquid fuel or fuel cell creates many mobile solutions across trains, planes and large marine transportation.

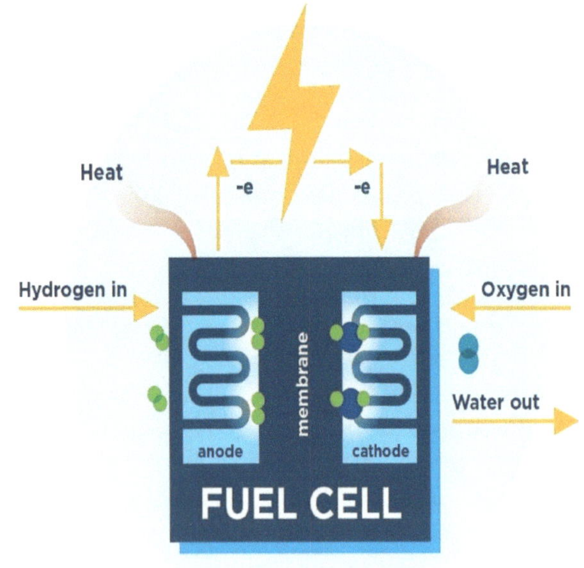


Emerald $H_2$ may serve as a feedstock for hundreds of manufacturing uses including glass, steel and petrochemicals. Food processing and the manufacture of other consumer goods consume considerable energy which can be supplied by clean Emerald $H_2$.

Emerald $H_2$ and Emerald ammonia are transportable and may be exported to international markets that demand clean energy.

Emerald $H_2$ offers a low cost, carbon negative clean energy to benefit all Ecolanda systems and people. Ecolanda's integrated systems allow something that has not been possible before, strong metrics assessing green solutions.

These eco-metrics are not available today because integrated biosolutions do not exist. SCAD will share the energy eco-metrics which will provide reliable comparisons across various energy solutions.

Ecolanda plans to produce enough Emerald $H_2$ to supply the entire Ecolanda campus, the local community and global exports. SCAD will build two large hydrogen energy ports where ships, planes, trains and other vehicles can refuel or transport Emerald $H_2$ and Emerald Ammonia.

## Aluminium Battery

*Clean, renewable energy provides a growth engine that needs stationary energy storage.*

Ecolanda operates exclusively on renewable energy which necessitates improved and lower cost electric energy storage.

Renewable energy varies with nature. Sunlight varies by season and may be occluded by clouds or fog. Winds vary daily as well as by season.

Renewable energy requires stationary storage in order to manage high and low demand as well as high and low supply. Energy storage needs flexibility because storage time varies from milliseconds to months.

Emerald $H_2$ and ammonia store electrochemical energy that can be used immediately or stored and shipped. These biosolutions provide liquid transportation fuels. Ecolanda also needs a stationary storage solution that holds electricity directly.

SCAD evaluated numerous novel material systems scouting for a reliable storage system aligned with Ecolanda goals. The system must be eco-friendly, adaptable and assures sustainable production. A battery that meets Ecolanda specifications must be disposable without hazardous materials and safe from over-heating and fire. The energy storage unit should have high energy density, fast recharge rate and long-term recharge reliability. It needs malleability for size and shape flexibility. The smart aluminum ion, Al-ion, battery fits the requirements.

Lithium and lead dominate battery markets today. Neither met SCAD's constraints. Lithium-ion batteries overcome some lead issues, but the supply of readily extractable lithium is extremely limited. Each year, the quality of the remaining extractable lithium diminishes and the ore less Li-rich. Higher mining costs drive up lithium prices.

Li-ion batteries overheat, catch fire and create a hazardous materials disposal problem. Li-ion batteries are complicated to make and require days for chemicals to dry and cure.

These batteries require an extra safety circuit to ensure they do not overheat and burn.

The smart aluminium-ion, Al-ion, battery uses metallic aluminium as the negative electrode. Al-ion delivers a volumetric capacity four times higher than lithium.

Aluminium reigns as the most abundant metal in the earth's crust, which adds sustainability and cost efficiency. Aluminium benefits from a 70-year global infrastructure for production, manufacturing and recycling.

Aluminium alloys are so strong they are used as structural and covering elements for jet aircraft.

Li-ion batteries increase the demand for lithium, cobalt, phosphorous and other metals used in battery. Cobalt and phosphorous are already classified as "critical" by the EU.[16] Li-ion batteries are banned from air transportation due to their propensity to catch fire.

### *Al-ion advantages*

A comparison table displays Al-ion advantages.

The Al-ion energy density advantage of 160% adds value as does saving 75% of the Li-ion production cost.[17] Aluminium carries a charge significantly better than lithium, since it is multivalent. This allows every ion to compensate for several electrons

| Battery Comparison | | |
|---|---|---|
| **Ion Battery Type** | **Lithium** | **Aluminum** |
| Cost | $8/kg | $2/kg |
| Material scarcity | High | Low |
| Production steps | 52 | 9 |
| Production time | 12 days | 1 day |
| Energy density | 1 | 1.6 |
| Recharge rate | 1 to 2 hours | 1 minute |
| Safety circuit | Yes | No |
| Toxic materials | Yes | No |
| Flammable | Yes | No |
| Hazardous disposal | Yes | No |
| Carbon footprint | High | Low |
| Ecological footprint | High | Low |

These quick-recharge batteries charge wirelessly. They charge more quickly due to higher conductivity efficiencies. They do not need an extra charging circuit because they are designed not overcharge or overheat.

SCAD's smart Al-ion battery uses an organic electrolyte, which improves safety in production and use. The Al-ion battery can be easily and safely recycled at the end of its life.

Li-ion batteries are two times heavier than Al-ion. When combined with diminished energy density, the same battery weight allows a vehicle with an Al-ion battery to travel 5 times further on each charge. Aluminium is both light and malleable and can be shaped to fit in nearly any space.

Ecolanda will use smart Al-ion batteries for nearly all purposes – large and small energy storage across the Ecolanda campus.

### *Smart Al-ion battery production*

SCAD Technologies will build a clean-tech Al-ion battery production facility in the smart ecocity innovation park. Plant scale is designed for 100 GWh production on 1,000 hectares.

Workforce is expected to be 1,500 to 2,000.

The Ecolanda plant will produce Al-ion batteries for 5 million cars and trucks and larger versions for 10 million homes and commercial buildings.

SCAD Technologies will create a R&D battery team to advance innovations in energy storage. Al-ion batteries have all the potential uses currently served by lead and Li-ion.

SCAD designs redundant systems for reliable production. Many of these systems are off-line most of the time. Al-ion batteries will ensure these systems are ready to perform when needed.

SCAD operates an advanced virtual power plant linked to a series of highly efficient local microgrids. These features assure consistent energy for each sector and the smart ecocity.

Producing renewable energy is critical but transmission and storage are equally important. Modern energy systems may lose up to 10% of their energy in transmission. Ecolanda avoids transmission losses with local production for a majority of renewable energy. Energy storage in commercial grade aluminium batteries improves energy preservation and availability.

### *Solar and storage microgrid*

SCAD plans to construct two €400 million solar and storage microgrid projects on two Ecolanda campuses.

Each system will include a 500MW solar farm and a 250MW Al-ion battery energy storage system (BESS). The 750MW BESS capacity will provide Ecolanda with full electrification and a state-owned utility with substantial electrical energy for regional customers.

The 500MW solar farm with local storage will eliminate 1.45 million metric tonnes (MMT) of carbon emissions the first year and every following year.

SCAD will site solar farms atop agri-industry buildings as well as non-arable land. SCAD will secure a 99-year lease on the land. The Power Purchase Agreement with the appropriate utility will be structured as a Private-Public Partnership.

The clean solar farm and Al-ion battery storage will demonstrate a cleaner, more cost-efficient and viable alternative to diesel or coal power. It is also part of its overall Agri – Energy – Waste integration.

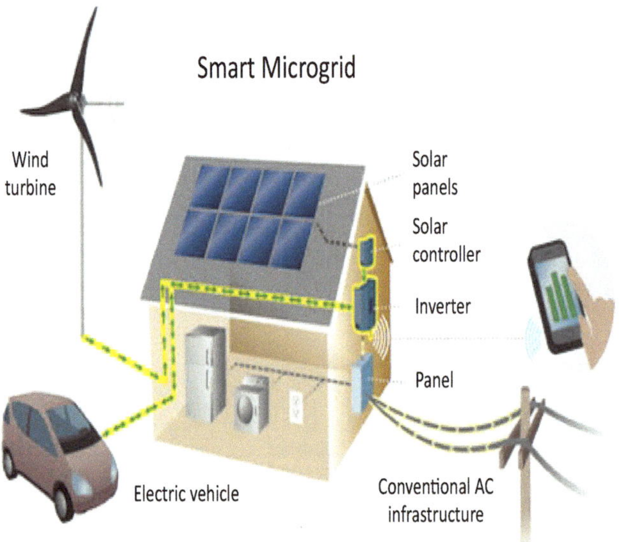


The solar and storage microgrid projects will assist the Nation to achieve its announced target of Zero Carbon Emissions by 2030.

Ecolanda will help secure the Nation's energy independence, provide long-term price stability and reduce reliance on fossil fuels.

## Bioenergy products

Drilling, extracting and refining industrial petroleum produces 1.7 gigatons of $CO_2$e per year.[18] Extraction emissions are amplified by flaring natural gas during oil recovery.

Extraction waste is extreme. Nigeria has lost over one trillion cubic feet of natural gas and $2 billion revenue between 2017 and 2020 due to burning gas through flaring.[19]

Most anthropogenic GHG emissions come from burning fossil fuels—coal, hydrocarbon gas liquids, natural gas and petroleum—for energy and for transportation, 37 gigatons of $CO_2$e.[20]

Power plants emit many harmful pollutants that lead to tens of thousands of premature deaths and disabilities each year. US power plants emit:

- 45,676 pounds of mercury, a toxic heavy metal that damages nervous, digestive, and immune systems and undermines child development.
- 3.1 million tons of $SO_2$. Sulfur dioxide form small, acidic particulates that can penetrate human lungs. It causes asthma, bronchitis, smog and acid rain. $SO_2$ damages crops and acidifies lakes and well-water.
- 1.5 million tons of $NO_x$. Nitrous oxides are visible as smog and irritate lung tissue, exacerbate asthma and make people more susceptible to chronic respiratory diseases like pneumonia and influenza.
- 200,000 tons of tiny airborne soot particles. Particulate matter in fossil fuel smoke causes chronic bronchitis, aggravates asthma, cardiovascular effects like heart attacks and premature death.
- 41 tons of lead, 9,332 pounds of cadmium and many other toxic heavy metals.
- 576,185 tons of carbon monoxide, which causes headaches and places additional stress on people with heart disease.

- 22,124 tons of volatile organic compounds, which form ozone.
- 77,108 pounds of arsenic. Arsenic causes cancer in one out of 100 people who drink water containing 50 parts per *billion*.[21]

Microfarms can harvest carbon in the form of atmospheric or stack gasses. Field research shows that algae can capture 90% of powerplant $CO_2$ flue gas, 85% of $NO_x$ and $SO_x$ while producing 50 metric tons of algae biomass per acre per year.[22]

The resulting biomass makes products like construction materials that ensure the poisons are held inert and harmless for generations.

Algae bioenergy production emits **zero** GHG. In contrast to fossil fuels, every ton of bioenergy production capture two tons of $CO_2$. Green bioenergy requires no drilling, deep-well extraction or gas flaring. Burning bioenergy emits GHG, which represents nearly net-zero emissions. Biofuels spew no black carbon particulates, no ozone compounds, $SO_2$, $NO_x$ and no poisonous heavy metals.

Algae biofuels are refined as a coproduct from several bioprocesses.

## Bioenergy co-bioproducts

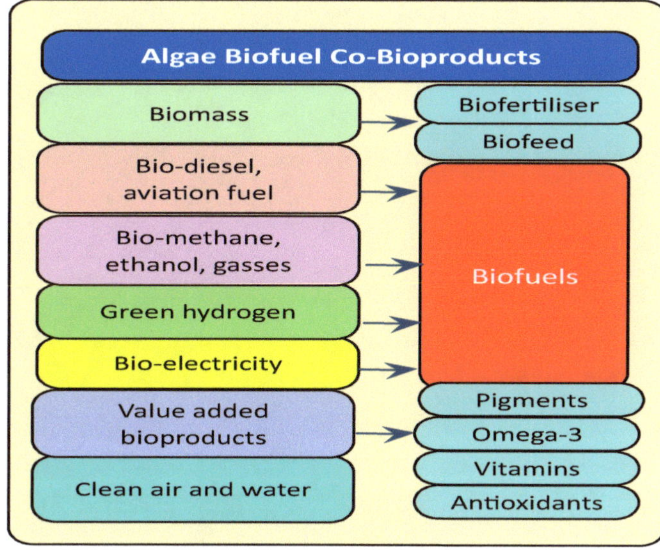


SCAD's bioenergy technologies can create any petroleum product, including aviation gasoline, asphalt, bioplastics, packaging and resins.

## *Virtual power plant*

Local smart microgrids receive their energy from both local renewable energy and a virtual renewable energy power plant. The schematic illustrates power plant elements.

The virtual power plant provides reliable power throughout the system. The plant does not suffer from the risk exposed by the 2021 ice storms in Texas that caused widespread power failure and loss of human and animal life.

The plant operates with important safeguards. Redundancy reduces the risk that any source will take down the entire grid.

In addition, several energy storage solutions, batteries, fuel cells and compressed pressure air stores excess energy for use when needed.

## *Virtual power plant capabilities*

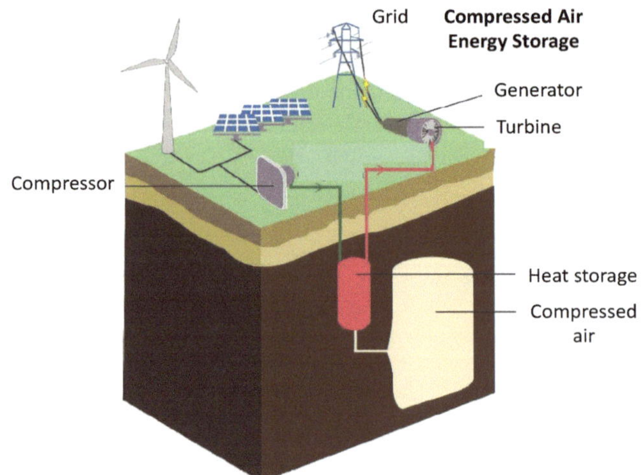


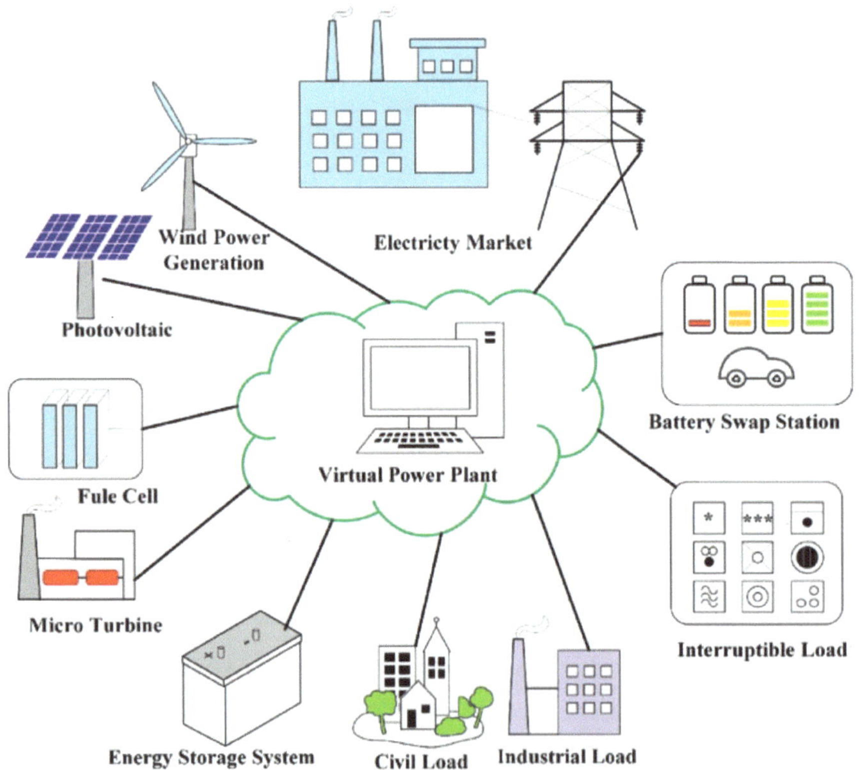


The virtual power plant offers substantial advantages.

- Reliable microgrid distribution
- Redundant systems add reliability
- Smart storage in an aluminium-ion battery
  - Minimum lifetime operational cost
  - Maximum management flexibility
  - Multifunctional and redundant systems
  - Residential, industrial and agri supply
  - Short and long-duration storage
  - Seamless energy and systems integration
  - LDS flexibility for regulatory oversight
  - Addresses commercial volatility

Excess energy capacity is shared with the regional community.

The next section addresses SCAD's cleantech strategies for water preservation and production.

# 8. Water Energy Nexus

*Water is energy intensive.*

*Energy is water intensive*

*Water is the driving force of all nature.*

*– Leonardo da Vinci*

Ecolanda treats water with the respect it deserves, as the soul of the earth. The lifecycle and the water cycle are one. Ecolanda growers and citizens treat water as our most precious natural resource.

SCAD treats water, the most critical natural resource, with a novel three-layer strategy.

1. **Save** – save water with advanced technologies and smart applications that use 80% or less water compared with industrial methods.
2. **Substitute** – save potable blue water by substituting non-potable water for extensive bioproduct cultivation.
3. **Sparkle** – clean waste and other non-potable water, recover and repurpose the nutrients and create sparkling blue water.

These three strategies drive all Ecolanda design decisions because water conservation success requires rigorous systems integration.

Intensive mechanical agriculture often uses 80 to 90% of a community's fresh blue water.[23] Much of the water goes onto cropland to produce animal feed. A single pound of beef uses 1800 gallons of water.[24] Most irrigation water flows onto cropland for animal feed or biofuels.

American farmers use nearly 3 trillion gallons of water annually to grow corn for biofuel production.[25] Most countries do not produce biofuels because lifecycle analysis shows corn ethanol production uses more energy than the fuel delivers.[26] Farmers in the US produce biofuels only because they receive substantial subsidies. Policy makers ignore the massive water loss.



Water management begins with saving water.

**Ecolanda Goal**: Reduce net water consumption by 80% compared with industrial agriculture and other social and economic development.

SCAD employs extensive sensors and monitors to manage and control water preservation in agri-energy, waste and the ecocity. Ecolanda field crop cultivation uses 80% less water than industrial agriculture production.

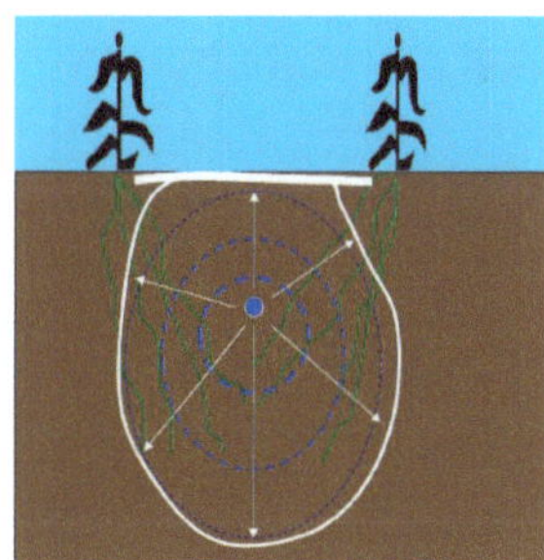

Field crops are irrigated by sub-surface drip. These systems save about 80% over surface irrigation. Field systems irrigation typically loses about 50% of the water from evaporation.

Sub-surface drip systems do a better job of spreading moisture throughout the rhizophore, root zone, and avoiding evaporation. Improved moisture dispersion increases root depth and strength, improving crop growth and vitality.

Cropland typically varies significantly in moisture retention according to the type of soil in each micro-geology. SCAD installs remote soil sensors approximately every six hectares to ensure all areas in a field receive precisely the water needed to support healthy crop growth.

Sub-surface drip irrigation avoids runoff due to careful moisture control. In areas where annual monsoons or typhoons are likely to overwhelm fields with driving rain, catchments reservoirs are built to save water for future use.

Ecolanda buildings are designed with catchments that vary from simple rain barrels to large storage tanks. Water-smart architecture employs a variety of retention technologies that save water that would otherwise become runoff.

SCAD uses Controlled Environmental Agriculture, CEA, extensively. Hydroponics and aeroponics save 80 to 90% of water compared with surface irrigation.

## Vertical farms

SCAD CEA in vertical farms save 90% of cropland, 90% of inorganic fertiliser and 95% or more of industrial pesticides and poisons. SCAD CEA allows no emissions or pollution of air, water, soils or ecosystems.

Field crop water loss occurs from about 50/50 evaporation from the soil and plant transpiration.

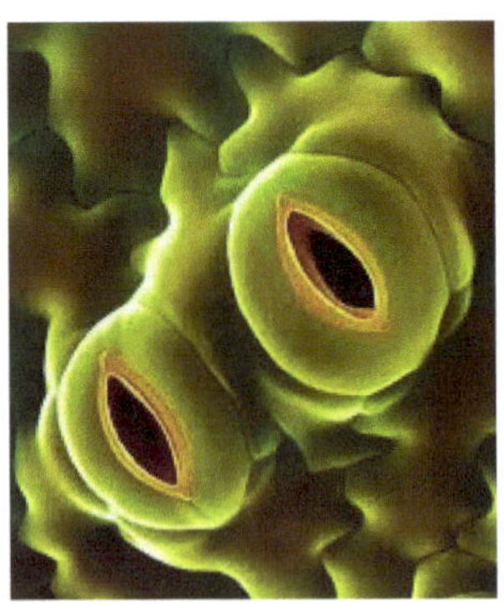

Transpiration keeps plants cool by allowing water droplets to exit through their stomas, left, on leaves and stems. Plants lose 90% of their water from transpiration.[27]

Ecolanda uses vertical farms to catch and reuse both evaporation and transpiration moisture. Moisture rises to the ceiling, which may be multiple stories high. Sensors identify when enough moisture has accumulated, which switches on a tiny pump that flows excess moisture to a storage catchment for reuse. Some vertical farms design water recovery by gravity feed.

SCAD uses CEA in vertical farms to produce 90% of aquaculture feed, which saves millions of litres of irrigation water that would be lost by field crops. Fish thrive on algae biofeed which offers superior nutralence and enhances fish growth and development. SCAD cultivates algae aquaculture feed using zero blue water.

Vertical farms are used to cultivate about 60% of the feed for meat and dairy animals. These animals need some ruffage for good digestion, which come primarily from field grains. The majority of animal nutrients are delivered by CEA feed crops such as alfalfa, barley shoots and high-nutralence algae biofeed.

CEA irrigation designed to biocycle water for recovery may reuse the same water multiple times for different crops.

## Zero-carbon vertical farm

Conventional greenhouses are not airtight and lose considerable energy and water. A biofoam greenhouse offers 95% less water loss, which allows maximum recovery from crop evaporation and transpiration.

SCAD is working on highly insulated vertical farm construction materials such as biofoam that make substantial savings to both energy and water. The vertical farms below illustrate space efficiency.

Construction-grade biofoam protects from

termites and other bugs, odours, sounds and water invasion. This biodegradable material saves 80% on utilities because heating and cooling energy stays within the building.

Biofoam offers other important qualities. It resists fire due to non-flammability.

Foam delivers twice the tensel strength of concrete block and three times the strength of wood construction. The material flexes under pressure, which makes it highly resistant to earthquakes and typhoons.

Ecolanda plans to use biofoam extensively for many types of buildings across the five sectors. SCAD makes the material from algae oils and fibres cultivated in algae biosystems. Similar biosystems create biodegradable bioplastics, biofilms, biopackaging and other construction materials.

How are Ecolanda biomaterials relevant to water? Finding substitutes for water-expensive materials saves enormous amounts of blue water.

Over 70% of people globally live in concrete structures.[28] Concrete consumes massive amounts of water and emits 5% of all GHG. Concrete is the second most consumed substance, after water.

Each cubic meter of concrete uses 160 Kg of water added to the Portland cement, sand and aggregate mixture. Extensive additional water is consumed in mining inputs and curing cement.

SCAD plans to create biodegradable building materials using zero potable water. Water savings will be enormous.

### Blue water substitutes

People, animals and most land crops require fresh water. Algae cultivation may use waste, brine, brackish or ocean water. Blue water substitutes save millions of litres of blue water for other purposes.

SCAD works with several universities that are developing a suite of feed and energy field crops that produce with saline water. These crops can grow in soil too salty for modern field grains.

Blue water saving and substitutes are excellent ways to conserve blue water. Ecolanda goes further and creates fresh, drinkable blue water.

### Sparkling blue water production

SCAD applies several technologies to transform waste, brine or brackish water to sparkling blue water. Technology selection depends on geography, latitude and the local hydrosphere.

**Ecolanda goal:** Create 20% more blue water than used in Ecolanda, including the farm and ecocity

Waste streams made from gases, water and biosolids enter the biosystem. Algae go to work using photosynthesis to power the carbon and nutrient capture. Algae grow and propagate using the waste stream nutrients until the water is sparkling clean. Some water systems require extra filtering to assure clean water.

Growers remove the rich organic biomass using filters, centrifuges or gravity settling. Algae biomass follows the path of petroleum refining and gets converted to multiple bioproducts. A coproduct of algae cultivation, sparkling blue water, benefits the community.

SCAD employs several biotechnologies to ensure pathogens are killed before they enter the BioRenew process. Sensors and monitors create continuous records of water quality.

Biological water remediation solutions are preferred because they use solar rather than mechanical energy. Biosystems biocycle the carbon and other nutrients, which are lost with mechanical desalinization systems.

Mechanical desalinization plants cost several hundred million dollars to build and millions a year to operate. Desal fails to capture the carbon or other nutrients.

Desal creates severe problems with dumping excessive residual salt. Many coastal estuaries and oceans that may seem ideal for desal contain too many toxic chemicals.

Desal concentrates the toxins, making residual disposal severely damaging to local ecosystems and sea life.

SCAD's BioRenew process tackles the bulk of water treatment as shown in the diagram.

The process of reclaiming water saves extra water. Growers recycle the growing media, the culture containing residual nutrients. New waste streams flow into the BioRenew process replacing the blue water that is reclaimed.

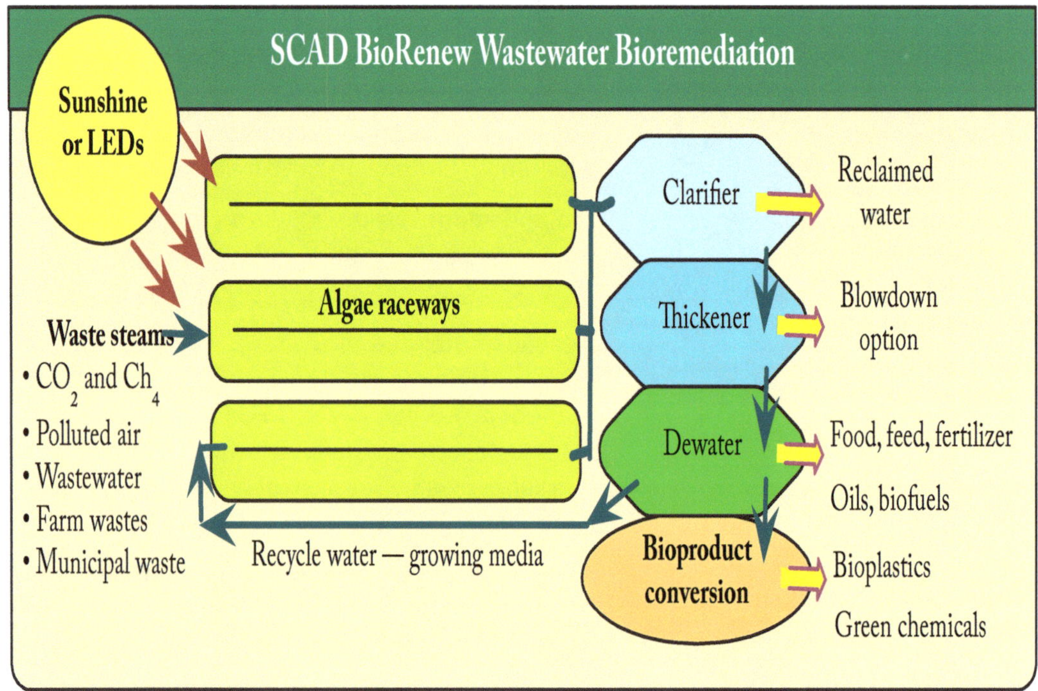


The BioRenew model shows sparkling blue water as a coproduct of bioproduct production. Open raceways in the diagram allow water evaporation. The same process may occur in insulated vertical farms that recapture and reuse water multiple times. Highly water efficient closed PBRs, below, may be used in place of raceways.

Outdoor raceways grow biomass only while the sun shines. Vertical farms and CEA using LED, laser or fibre optic light may extend production up to about 16 hours a day. Algae cells are work horses, but they are also living organisms and need a rest daily.

The next section examines Ecolanda's smart ecocity.

# 9. Ecolanda's Smart Ecocity



**Goal**: Ecolanda's smart ecocities inspire awe and joy and deliver economic growth, health and happiness for residents.

Ecolanda smart ecocities are designed from their foundation to function as bioeconomies with a robust, sustainable circular economy. Thoughtful architecture and engineering design with systems integration in mind to use entirely green energy and create zero emissions or pollution.

Ecolanda ecocities operate with no or minimal consumption of fossil resources, capture rather than emit carbon and produce 20% extra blue water. Smart land use assures plenty of green space, recreational parks and botanical gardens.

Smart cities accrue social and economic benefits from positive economic growth and lower costs from using resources efficiently. Automation improves reliability and speed while it saves costs.

Intelligent systems and the internet of things, IoT technologies, automate city resources such as communications, power, transportation and water. Smart, connected buildings save resources, improve safety while generating their own renewable energy. Connection improves reliability, performance and allows effective, data-driven decisions.

Smart transportation reduces congestion and pollution, improves safety and saves citizen's time. Sensors monitor utilities and events. They mitigate risks and reduce damage from extreme events. Sustainable ecosystems with zero emissions create a clean city, which increases the standard of living and happiness. These actions lead to economic growth.

## *Ecocity model*

Ecolanda ecocities model the self-sustaining resilient structure and function of natural ecosystems. Ecocity architecture design, below, allows citizens to live comfortably within environmental means. Ecocities eliminate carbon and other wastes and use exclusively renewable energy and resources.

Ecolanda's ecocity stimulates economic growth, reduces hunger and poverty and improves health and vitality for residents. Each ecocity is unique based on geography, history and culture.

Ecocities strive for efficient land use, green public transportation, efficient use of resources and habitat preservation and restoration.

Happy citizens care about their city and work to make additional improvements. Talent flows into smart cities attracted by the vitality and healthy environment. Happy residents enthusiastically drive the economy.

The ecocity houses Ecolanda associates and their families. Residents include others who desire to live, work and play in the healthy, safe and vibrant community. The ecocity adopts best practices and learning from ecocities currently in designing, planning, building and operations.

SCAD has assembled a group of world-class architects to design each Ecolanda smart ecocity. Architectural designs include inputs from local, regional and national stakeholders. Participation ensures designs reflect the community's vision and values. Architectural design aligns with national and local culture, history and art.

## Ecocity benefits

The smart ecocity benefits from the attributes described in the graphic below.

## SCAD Education and Training

SCAD Ecolanda has a strong commitment to early education and life-long learning.

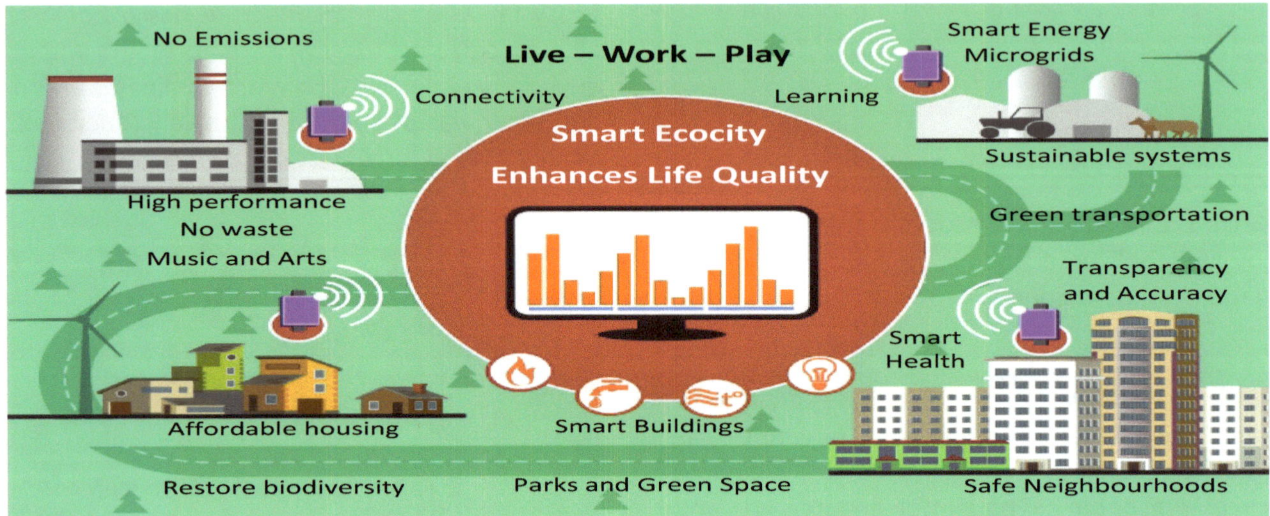


SCAD pursues an 80% goal for local production of food, energy and blue water used in the ecocity. The blue water comes from ecocity wastewater cleaned and biocycled to clean blue water.

The ecocity offers several novel features besides affordable housing and green transportation. SCAD plans an international sports training facility for young aspiring athletes. Several medical and health facilities will draw people who want to improve their diet, exercise and lifestyles. Abundant green space engages residents in recreation and helps restore biodiversity.

Ecocity infrastructure supports education, social, family, recreation and religious needs. Best practices guide social, community, environmental and e-governance. The community publishes its shared values to create a sense of belonging for everyone.

Education and training options offer both in-person and distance learning options. Staff provide skills training for thousands of Ecolanda associates. These include agri-energy-water operational and experiential courses. Staff deliver education on basic, intermediate and advanced sustainable earth-friendly technologies as well as sustainable lifestyles.

SCAD plans include building and staffing a pre-school, primary and high school in the ecocity. Staff partner with appropriate stakeholders in building school curriculums, including for colleges and universities. Staff also partner to assist with building a health, safety and sustainable lifestyle curriculum.

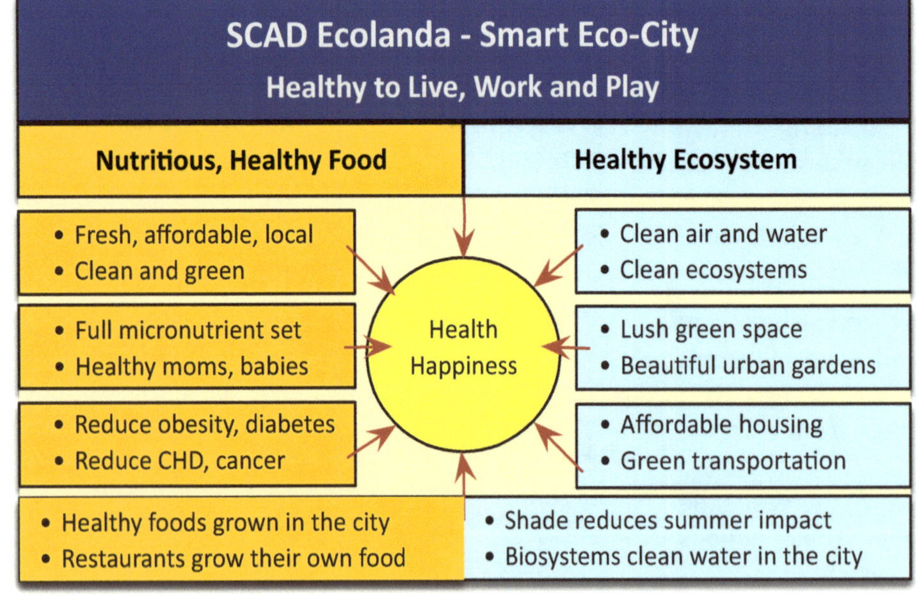

## Eco-tourism

The Ecolanda agri-energy farm includes an excellent training facility for associates and local people with a desire to learn about sustainable agri-energy. The learning centre will become an eco-tourism attraction for people globally who wish to learn biosolutions and how to build and operate fascinating sustainable technologies.

*Sustainable agriculture learning centre*

The learning centre will introduce new crops and better ways of cultivating crops. Emphasis on practical learning ensures that students are able to take away capabilities for their own farms, homes, apartments and urban gardens.

Microcrop cultivation requires considerable attention and knowledge. SCAD will create an online network to assist growers remotely using their cell phone or PC.

Sensors will transmit daily culture data from the grower to a central support centre. A team of SCAD experts, academics and volunteers will answer questions. The team may recommend specific strategies to avoid pests and to successfully grow microcrops. The team will also assist with down-stream processing and developing markets for local bioproducts.

Indigenous agri-techniques offer value. A cross-functional team will learn agri-methods that have been used for centuries from indigenous people. The team will create a curriculum set for local people to learn modern sustainable agricultural techniques they can practice on their farms and communities.

## Entrepreneurial Park

The Ecolanda community benefits from a diverse entrepreneurial park. The park includes a technology incubator that mentors and launches vibrant companies that take novel Ecolanda technologies to the next level. Focus areas are shown in the graphic.

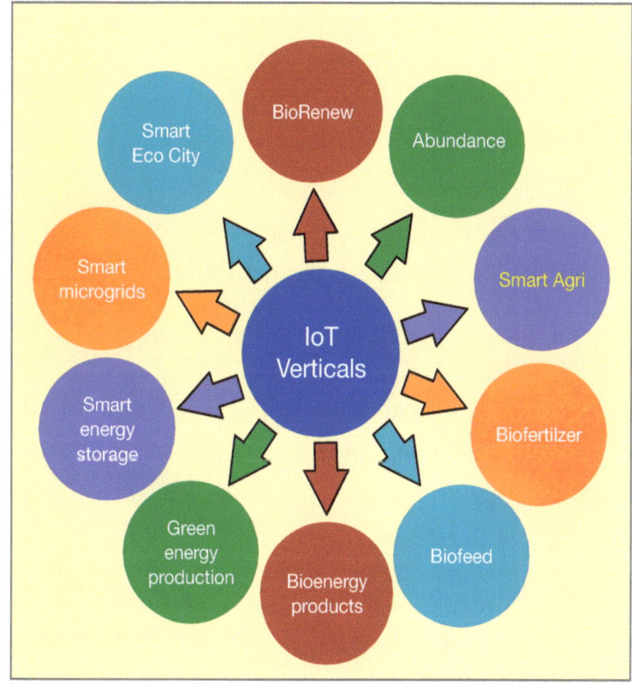


Ecolanda advanced technologies are used across the entire project and include:

- Internet of things - serves as SCAD's collection, monitoring and data first-line data analysis tool.
- BioRenew - biocycles carbon and other nutrients for economics and ecology.
- Abundance agricultural methods - applies BioRenew to preserve natural resources and avoid overconsumption and waste.
- Smart agri - precision agriculture with smart sensors, monitors, data capture and analysis.
- Biofertiliser - biocycles waste nutrients from air, gas, water and biosolids into clean, high nutralence plant food.
- Biofeed - biocycles waste nutrients to recover, recycle and repurpose nutrients into clean, highly nutritious animal feed.

- Bioenergy products - coal dust bonded to algae; briquettes burn hot and clean.
- Bioenergy storage - aluminium ion battery, green hydrogen and fuel cells.
- Emerald $H_2$ and Emerald Ammonia production - energy products biocycled from waste streams.
- Smart energy storage - local energy storage in aluminium batteries.
- Smart microgrids – green energy produced and stored locally avoids transmission loss.
- Smart eco-city - connected smart city produces its own food, energy and water locally and avoids waste and pollution.

The ecopark creates new jobs and strengthens the local, regional and national economy. SCAD devotes funds and advisors to indigenous people who want to build-out ideas to benefit their communities.

The Ecolanda entrepreneurial park operates as a public—private partnership. Distance education and skills training are available for people at all levels. Student have access to internships, mentors and special projects.

Robust internal and external vendor relationships support new ventures with grants-in-kind, new technologies, and experienced coaching. Multi-disciplinary teams elevate new project and assist with the soft and hard launch.

Integrated sensors, monitors, metrics and reporting using dashboards provide insight. These continuous updates and connections support learning and moderate risk. The launch-pad maintains strong ties and engagement with national and global academics.

Smart methods such as simulations and computer modelling are used for prototypes. These actions save considerable time and cost. Guidance for market testing assists potential entrepreneurs.

### Ecopark new product examples

Many new Ecolanda technologies or applications will find hungry markets for novel products and processes that support bioeconomies. Each vertical sector offers extensive opportunities for new product development.

The agri sector offers golden opportunities for new foods, ingredients, functional foods, health, nutraceuticals, cosmeceuticals and medicines. Partnerships with local and national colleges and universities will strengthen the creation and enhancement of novel products.

Emphasis on plant-based meat and dairy products will find champions. Industrial meat and dairy consume massive amounts of natural resources and severely contaminate the environment.

Over 80% of medicines today contain extracts and compounds from plants and animals. Since terrestrial plants evolved from algae, nearly all

| Ecolanda Ecocity Entrepreneurial Park | |
|---|---|
| Novel batteries | LED lighting |
| Robotic systems | 3D printing |
| Green energy systems | Smart cities |
| Smart sensors and big data | Market research |
| IoT and advanced analytics | Smart transportation |
| Ecology and econometrics | Sustainable systems |
| Abundance growing methods | Aquaculture methods |
| Renewable building materials | New food development |
| Renewable energy development | Hydroponics production |
| Zero emissions waste management | Advanced agri-production |

the target compounds needed for medicines can be cultivated in algae.

Plants and animals grow extremely slowly compared with algae. Extracts are expensive, difficult and often contain dangerous contaminants. Microcrop medical compounds may be cultivated significantly faster than land-based plants. Cultivation for pharmaceuticals, nutraceuticals and medicines occurs in clean rooms, which eliminate most contaminants.

Ecolanda's plant-based foods are cultivated as freedom foods, the next section's topic.

# 10. Freedom Foods

*Make a habit of two things: to help; or at least to do no harm.*  — Hippocrates

Freedom foods help people, animals and plants and do no harm to the environment. These foods actually enhance ecosystems.

Freedom foods are cultivated with abundance methods that preserve non-renewable resources for future generations by avoiding extraction.

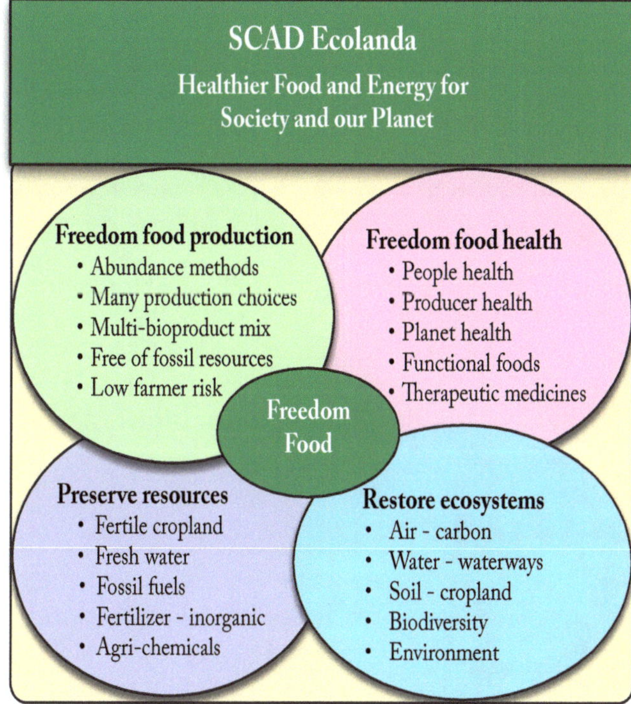



Freedom foods are special foods made from algae and other microcrops. They are extraordinary because they are the first food designed to grow free from consumption of fossil resources and free from waste and pollution.

Freedom foods are unique in that they consume zero fertile cropland, fresh blue water, fossil fuels, inorganic mined fertiliser, pesticides or other agri-poisons. Every ton of freedom food captures and repurposes two tons of $CO_2$.

Freedom foods give consumers a free choice to choose healthier foods for themselves and their families as well as their environment.

## *Higher nutralence*

Freedom foods deliver higher nutralence which enhances health for consumers whether they be humans, animals or plants.

Micronutrients are essential to sustain life, cognitive and physiological function. Globally, micronutrient deficiencies, MNDs are pervasive. In America, 85% of citizens of all ages suffer from micronutrient deficiencies.[29]

Pregnant women and their children under 5 years are at highest risk. MNDs cause cruelly weak physical development, intellectual, respiratory, vision impairments and perinatal complications. Iron, iodine, folate, vitamin A, and zinc deficiencies are extremely widespread.

Sperm counts have declined 59% over the last 40 years for men in North America, Europe and Australia.[30] MNDs are the most likely culprit.

MNDs put people of all ages at significantly increased risk of disease, premature disability and death. The elderly too often suffer from premature declines in cognitive, neurological, heart, lung and vision faculties due to MNDs.

Freedom foods can end MNDs in a few weeks. Each bite delivers a full set of micronutrients. Biocycled nutrients provide superior nutrition, micronutrients, bioactive compounds, vitamins, minerals and trace elements.[31] Their high nutralence delivers twice the protein and four times the total nutrients per bite of industrial foods. Freedom foods offer 20 times more natural biodiversity than legacy foods.

Most freedom foods contain dozens of bioactive compounds that are not available in legacy foods. Bioactive compounds protect against disease and fight diseases when they attack.

Freedom foods are composed of microcrops like algae with tiny cells. "Small becomes large" when measuring bioavailability. Bioavailability provides a metric for gut assimilation. These foods supply dozens of healthy food ingredients that are used to fortify and improve legacy foods. Nutrients go into functional foods that function to enhance health and vitality.

## Fossil and Freedom Foods

**Freedom foods**

**Microcrops**
- Rootless
- Microorganisms
- Weather independence
- Fossil independence

**Minicrops**
- Vegetables and herbs
- Fruits and nuts
- Aquaculture, hydroponics
- Low fossil dependence
- Grown in vertical farms

**Fossil foods**

**Fossil foods**
- Field crops with roots
- GMO monocultures
- Animal meat & dairy
- Industrial agriculture
- Fossil resource dependent

Fossil foods depend on fossil natural resources for production. They are typically grown in GMO monocultures, (genetically modified organisms.)

A Fossil Food Diet imposes health liabilities.

| Western Diet of Processed Fossil Foods Manufactured GMO Foods with Hidden Hunger | | |
|---|---|---|
| **Processed foods** | **Food types** | **Chronically low** |
| • High fat | • Highly processed | • Low fiber |
| • High cholesterol | • High animal dairy | • Low nutrient quality |
| • Medium protein | • High animal meat | • Low nutrient density |
| • High sugar | • High GMO | • Low bioavailability |
| • High salt | • Pesticide residuals | • Zero bioactive compounds |

Contrast Fossil Foods with Freedom Foods.

| Freedom Foods Diet Whole, High-Nutralence Natural Foods | | |
|---|---|---|
| **Whole foods** | **Food types** | **High nutralence** |
| • Low fat | • Whole foods | • High fiber |
| • Low cholesterol | • Plant-based dairy | • High quality |
| • High protein | • Plant-based meat | • High nutrient density |
| • Low sugar | • No GMO | • High bioavailability |
| • Low salt | • No pesticide residuals | • High bioactive compounds |

GMO monocultures means that the plants are essentially clones of one another. They all share the same strengths and weakness.

Monocultures put an entire crop at risk from a single pest vector. Monocultures put farmers at risk as they must use specific herbicides, pesticides and other poisons. GMO crops impose extra costs on farmers as these crops require more cultivation, water, fertiliser and agri-poisons.

The Freedom Foods Diet offers significant health benefits beyond higher nutralence. Freedom foods made from microorganisms do not impose the high fat and cholesterol common in food made from field grains.

Rice, wheat, maize and other grains taste bitter and bland. Food processors add extensive sugar and salt. Freedom foods are tasty without adding extra sugar and salt.

Plant-based meat and dairy freedom foods avoid the curse of fossil meats, high fat and cholesterol. Freedom foods use no GMO, which avoids the comprehensive concerns about GMO crops and health for growers, consumers and the environment.

Fossil foods carry pesticide contaminants which can damage the brains and major organs of unborn and newborn children. Freedom food growers use no agri-poisons, which zeros-out pesticide residuals on and in food.

Many freedom foods are high in fiber, which helps with digestion and gut health. Microcrop meat products may deliver twice the protein per bite as fossil meats. Freedom foods deliver over five times more nutrients per bite and 10 times more nutrient diversity.

### Therapeutics

Many freedom foods deliver nutraceuticals and therapeutics that treat or avoid disease vectors. These foods deliver superior aroma, color, texture, and taste. They have less than half the fat and cholesterol than field grain foods.

Abundance growing methods reduce production risk for growers while diminishing waste and costs. Growers have no risky exposure to heavy farm equipment, diesel fumes, dust or agricultural poisons. Rural communities are spared from agri-emissions and pollution.

Microfarms can provide climate independent food production year-round in a city, even after a natural disaster.[32]

### Nosh

**Nosh** refers to eating greedily. Nosh has the connotation of eating on the sly, between meals, or possibly sweets in secret. Eating foods with empty calories leaves the brain with no satiety, the feeling of fullness. The old Cracker Jack box expressed nosh perfectly, "The more you eat, the more you want!"

This was a successful tagline when most people were hungry and thin. Borden's and then Frito-Lay had to change their advertising for modern obese children.

Obese children listen to their nosh signal that tells them to eat more food, more often. If the food contains empty calories, their little brains continue to signal: "More, more, more."

Rimonabant or possibly other drugs can help children, but the brain signal side-effects are far too dangerous, except for morbid obesity.[33] Algae food may provide a safe therapeutic solution.

Algae compounds provide an array of medical benefits for children plagued with nosh signals that lead to obesity and diabetes. Two unique strategies may be called fill-gut and gut-full signaling.

The **fill-gut** strategy adds a few grams of dried algae eaten early in a normal meal, possibly as sprinkles on a salad. The algae expand and fills the stomach.

Alginates can absorb 300 times their weight in water, which quickly fill the gut and suppresses appetite by sending the gut full signal to the brain.

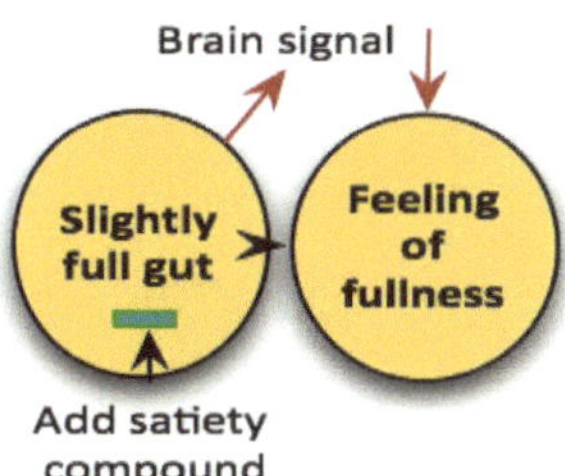


The **gut-full** signals work because algae compounds activate the stomach's natural satiety signals.[34] Satiety signals an immediate feeling of fullness, which tells the eater to stop eating naturally. Algae satiety signals the brain and quashes the nosh feeling.

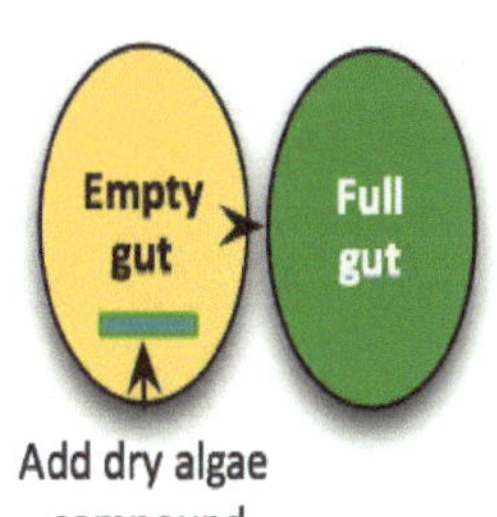


The nosh feeling hits children especially hard, which is why so many children are overweight.

Many types of algae contain alginates that quash the nosh directive. These foods expand in the gut creating a feeling of fullness that blocks nosh. Brain signals help young people and adults stay at healthy weight.

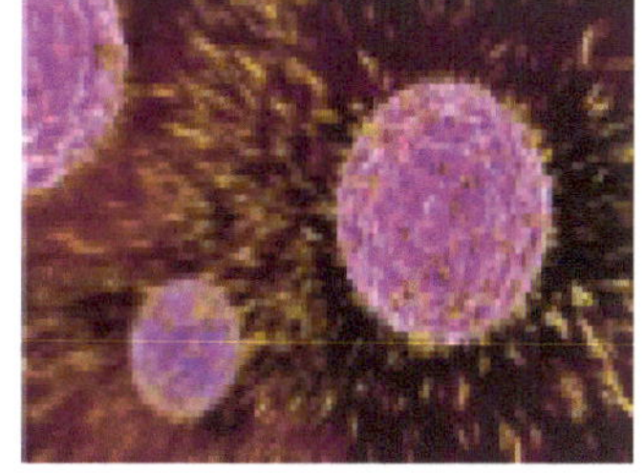

**Algae antivirals** use ingenious strategies attack and defeat viruses.

Some compounds make cell walls like Teflon, so the viruses cannot stick and do their damage. Others guard the DNA and RNA, protecting the cell signaling from infection.

Algae compounds block viruses from releasing genetic material or create disruption. Another algae compound strategy inhibits the virus from reproduction. Algae biostimulate the cell to produce macrophages that engulf the virus. The body senses the surrounded virus as a waste material and sluffs it off naturally from the body.

### Functional nutrition

Algae biomass produces proteins, lipids, carbohydrates and many specialty compounds. Proteins are large organic compounds made of amino acids arranged in a linear chain connected by peptide bonds. The plant's genetic code determines the sequence of the amino acids, but nutrient limitations may cause changes to the production of amino acids.

Algae's vast biodiversity offer excellent potential for the cultivation of high valuable molecules. Most of these are not easily produced in land plants. Many high- value recombinant proteins are therapeutic proteins. Recombinant proteins may be used to fortify nutritional value.

Recombinant proteins often retain their biological activity when they are ingested, thus improving human health. These functional proteins like the colostrum protein osteopontin, are naturally present in breast milk. They have been shown to impact brain development and immune system function.

Another functional protein found in breast milk, Immunoglobin A, is an antibody found in most body secretions due to is antimicrobial activity.

Most proteins are enzymes that catalyze biochemical reactions and plant metabolism. Other proteins maintain cell shape and provide signaling functions.

Protein plays a crucial role in the human diet, providing most of the nitrogen humans need. Proteins deliver a subset of amino acids that cannot be synthesized by the human body and those need to be supplied in the diet. These essential amino acids are histidine, isoleucine, leucine, lysine, methionine, phenylalanine, threonine, tryptophan and valine.

Other "conditionally essential" amino acids may not be synthesized properly by the body. These include arginine, cysteine, glutamine, glycine, proline and tyrosine. Global hunger is characterized by Protein-Energy Malnutrition (PEM). Deficient intake of essential amino acids reduces total energy and causes a series of extremely dangerous conditions.

PEM can be solved with a new source of inexpensive and balanced protein from algae and other microcrops.

Some algae have a very high percentage of their dry biomass as protein. Species like *Arthrospira platensis, spirulina,* have up to 70% of their biomass as protein content. A short 8-week course of 3g a day of spirulina have been shown to eliminate PEM and other micronutrient deficiencies in children and adults.

Lipids are long carbon chain molecules. Lipids store energy for the plant and serve as the structural components for cell membranes. Lipids are an indispensable component of cells and are precursors of many essential molecules. Lipids are crucial for the human diet.

Some algae accumulate lipids to 70% of the dry biomass. Similar to essential amino acids, some lipids are essential, including a-Linolenic acid and Linoleic acid. Several algae lipids have proven to have a positive impact on human heart, brain and circulatory systems. Long chained omega-3 fatty acids docosahexaenoic acid (DHA) and eicosapentaenoic acid (EPA) are not naturally synthesized in animals or land plants. They must be acquired through diet.

The traditional source of such nutrients in human diets has been seafood in general. Fish contain omega-3 fatty acids because they consume plankton and algae as part of their diet. The essential long chain polyunsaturated fatty acids are produced in algae.

Many fish stocks have been over-harvested to remove the omega-3s. The daily omega-3 dose for humans may consume 16 sardines or other small oily fish. Loss of small fish stocks creates a cascade of hunger in the ocean which drives down large fish growth and numbers. Consuming algae-based omega-3 fatty acids offer a healthy alternative for humans and fish.

Algae provide an excellent source of vitamin A, vitamin B complex and vitamin E. Other algae nutrients that have a positive impact in human health are antioxidants lycopene, b-carotene, and astaxanthin and polysaccharides, beta-glucans. Beta glucans are a soluble dietary fiber that is strongly linked to improving cholesterol levels and boosting heart health.

Algae pigments are different from land plant and synthetic colorants because they provide nutrition. B-carotene is transformed in the human body into the essential vitamin-A. Astaxanthin, also a carotenoid, has a distinctive red color used in animal feeds to confer a deep yellow color in chicken egg yolks and the red color used in farmed-raised salmon. Farm-raised salmon fed field grains are not marketable because they lack the pink salmon color.

Many algae species have therapeutic properties that improve health and prevent or treat disease. Algae bioactive compounds have proven benefits against degenerative metabolic disorders.

Metabolic syndrome, which is highly correlated with the Western Diet of fossil-based foods, imposes severe health conditions throughout life. The metabolic syndrome, which is very common globally, can be avoided with a diet that includes microcrops.

## Bioactive compounds

Freedom food nutrition offers over 300 bioactive compounds that provide organisms with various types of shields against disease vectors. Industrial agriculture does not deliver bioactive compounds in produce. Bioactive compounds developed in algae-based foods over their 3.5 billion years of evolution. Land plants evolved from algae about 500 million years ago.





The move to roots, stems and terrestrial stressors forced land plants to leave behind any "excess" baggage. Most bioactive compounds either do not exist or are in such sparse amounts they are not measurable in land plants.

A land plant such as maize evolves very slowly because it produces only one crop a year. This means any mutations or even hybridization trials require a decade or more.

Microcrops like algae create a new crop of offspring daily, every day the plant has sufficient sunshine, moisture and nutrients.

Algae, including macro (seaweed), micro (green) and cyanobacteria (blue green), had to survive in extremely rigorous early-earth conditions. Algae form the foundation of the food chain which means that every consumer higher on the food chain viewed algae as food. Algae created a brilliant strategy to contend with all these ravenous consumers – reproduce faster than consumers can eat.

Algae survived every threat vector known as human, animal and plant diseases. These wise plants developed bioactive compounds to neutralize disease threats. The beauty of algae-based food, feed and biofertiliser is that when consumers assimilate the superior nutralence, they receive bioactive compound protection.

Nrich infuses functional foods naturally with an extraordinary set of macro and micronutrients, plus vitamins, minerals and trace elements that are not available in industrial foods. Nrich functional foods provide natural solutions that maximize health and vitality.

Industrial production of fibers and pigments creates massive pollution two air, water and soils. Freedom foods produce strong fibers and more vibrant pigments with zero emissions and zero pollution.

## Multiproduct production

Industrial farmers produce a single monocrop commodity. The staple crop makes farm revenue dependent on the commodity price, over which famers have no control.

A maize farmer takes extreme risk as all the crop inputs must be bought and applied to the crop before the maize can be sold. The farmer's risk includes a 10 to 20% chance the crop will fail, and the investment lost. Global warming increases the risk of total or partial crop failure. The maize and sells at the single commodity price, which may be higher than the farmer's inputs.

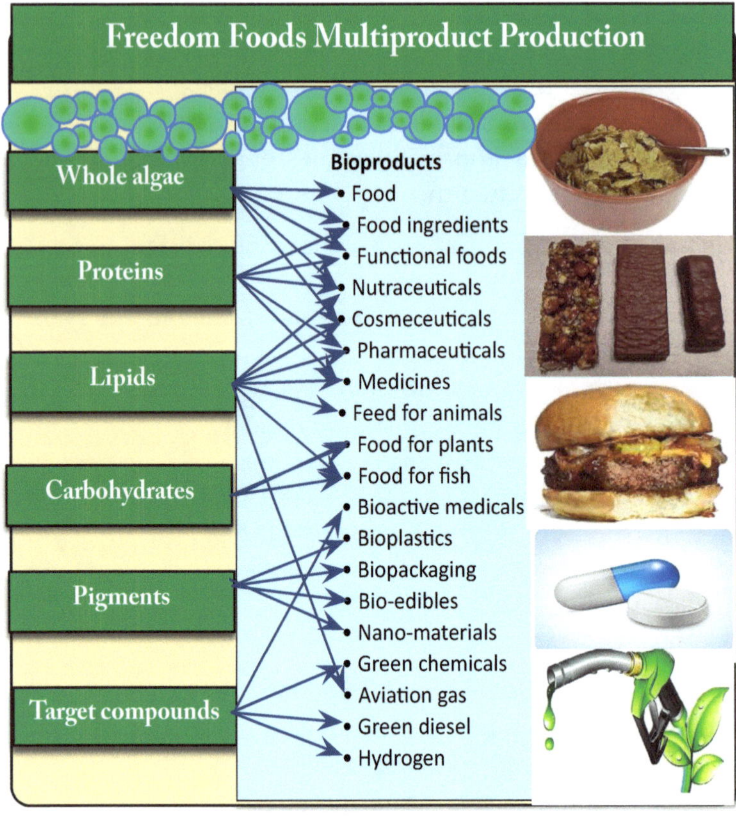


Freedom foods growers do not have the high outlay to pay for crop inputs. They do not have to buy the expensive fossil resources. Their input costs are about 70% less than a field grain farmer.

The maize farmer must wait 120 days for the crop to mature for harvest. Freedom foods growers harvest daily or several times a week. This substantially reduces risk of crop failure. If a grower experiences a culture failure, a new culture can grow and be productive in 2 weeks.

Freedom food growers use multiproduct production which provides extraordinary value. Freedom food crops may grow continuously or in batch cultures, depending on the production strategy. In both methods, the rich biomass contains proteins, lipids (oils), carbohydrates, specialty target compounds.

Growers select from among many diverse species to optimize the value of their multiple bioproducts. Growers are always aware or the Emerald Flow and climate care.

# 11. Emerald Flow – Climate Care

*No matter how rich you are, you can't get healthy air.* – **Ma Jun**

Industrial emissions spread carbon, black soot particulates and other toxins across the entire community. SCAD's Ecolanda commitment:

- Zero emissions from Ecolanda
- Clean GHG and other contaminants from industrial farms equal to 3 times the Ecolanda campus area

Ecolanda designs solutions to reverse climate change. After water, Ecolanda growers and citizens treat our shared atmosphere as our most precious natural resource.

SCAD's emissions management involves a four-layer strategy.

1. **Emerald flow** – reverses carbon flows in our food supply chain from carbon discharge to carbon capture. Abundance bio-cultivation methods allow carbon to flow into bioproducts rather than out from industrial crops into the atmosphere. This may sound too simple, but the direction of carbon flow may be nature's miracle that can reverse global warming.
2. **Capture emissions** – save the atmosphere with zero carbon emissions from the Ecolanda campus.
3. **Clean air** – clean air with renewable energy, biosystems and direct air capture.
4. **Smart energy products** – invent novel products that replace coal with no black soot particulate pollution.

Each strategy contributes to the promise of clean, fresh air. Emerald Flow differentiates Ecolanda from other development models. The carbon flow from fossil foods flows from food production into our atmosphere. Abundance methods reverse the carbon flow from the air into foods and other bioproducts.

## Fossil food flow

How did nature create plentiful food and energy for over 2 billion years consistently and not pollute the atmosphere?

Nature has done her job extremely well. Humans have not. People began fouling our atmosphere 180 years ago with the start of the industrial revolution. Farmers accelerated GHG production in 1950 with the "Green Revolution." Both revolutions were black. They added billions of tons of black carbon to the air we share.

Industrial production applies physical power to force nature to do the producer's will. Physical energy requires mechanical means and all the fossil fuels needed to power heavy equipment. Fossil food production leads to huge carbon waste flows. Fossil foods create a toxic pollution plume that poisons our atmosphere, waterways and rural communities. The enormous fossil food pollution plume is composed of natural resource waste.

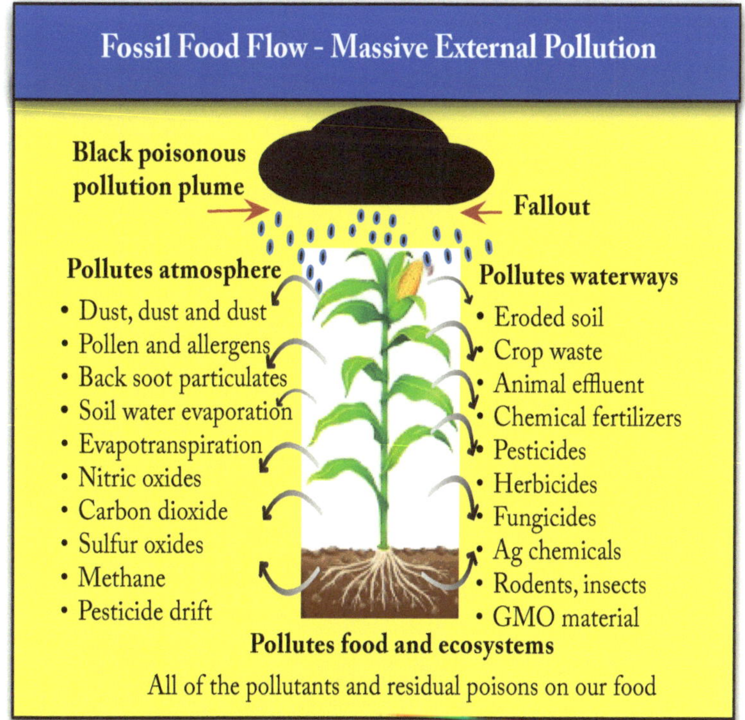


The costly discards and massive pollution may be the fatal flaw in industrial agri-energy production. Every year industrial pollution increases, degrading human and ecological health.

## Emerald Flow

Biological systems reverse the direction of the carbon flow from out to in. The Emerald flow transfers $CO_2$ in the air to fresh bioproducts.

Abundance agri-energy methods can use the substantial waste created by industrial production and biocycle those wastes into clean food and other useful bioproducts.

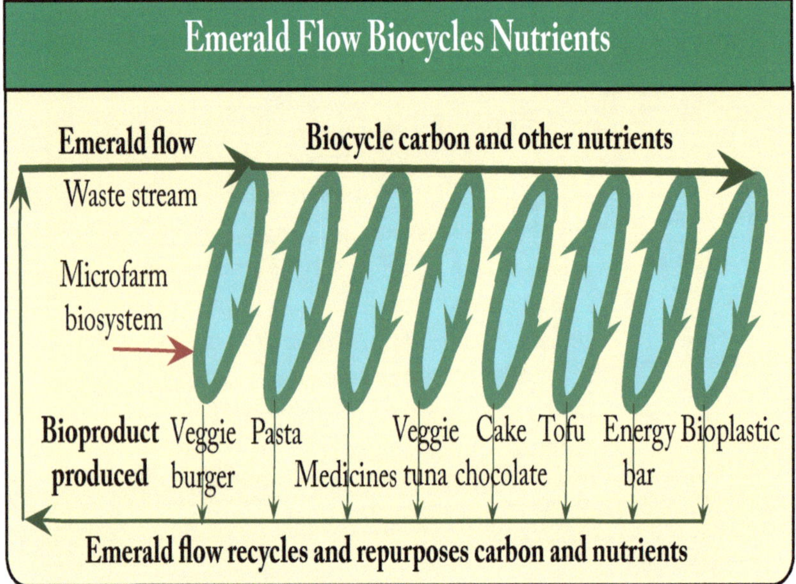


Ecolanda plans to grow a million metric tons of algae annually. This biomass will capture and reuse two million metric tons of carbon. As Ecolanda microcrop production increases, more carbon will flow into bioproducts.

The Emerald flow recycles and repurposes waste carbon and nutrients in microfarm biosystems. Microcrops produce the compounds to make thousands of bioproducts. Nearly any industrial product made today can be made cleaner and faster with Emerald flow in biosystems.

The Emerald Flow works in all types of microfarms, outdoor raceways or in indoor PBRs, right. Healthy algae grow and renew their biomass daily and display the magnificent colour of emeralds.

## Emerald's meaning

Many great civilizations have endowed emeralds with robust meaning, valuing emeralds for nature's power for rebirth and renewal.

**Indian mythology**, Sanskrit, the beautiful green of Spring and growing things.

**Ancient China**, Feng Shui, the energy of growth, new beginnings.

**Persia**, "smaragdus," the earth's rhythm of rebirth and renewal.

**Ancient Egypt**, the energy of nature and of eternal life.

**China**, living a good life, happily in harmony with nature.

**Aztec and Incas**, wealth, prosperity, rebirth and renewal.

**Islam**, verdant, inspires happiness and assures love from Allah.

**Middle Ages clergy**, nature's renewal and symbol of divine faith.

The Emerald flow aligns with the emerald's historical meaning of renewal.

Global adoption of abundance agri-energy methods would allow the Emerald flow to reverse climate change in one generation.

## Carbon capture

Ecolanda's nature-inspired architecture uses link and sync carbon capture and utilization, CCU, systems. Purposeful design allows each potential carbon waste source to be linked to a microcrop biosystem. Biosystems use the Emerald flow in microfarms to transform the potential pollutant to useful green biomass.

*Photosynthesis is simultaneously the cheapest and most efficient known solution of all carbon and nutrient cycling technologies.*

SCAD's commitment to CCU includes capturing gasses, liquids and biosolids from animal barns.

### Clean air

Ecolanda's fresh air systems capture, recycle and reuse airborne carbon. Biosystems remove and replace the dirty carbon, soot particulates and other toxic GHG with pure oxygen.

Clean air technologies make a first pass at cleaning the air by avoiding emissions and linking smokestacks and obvious carbon sources with biosystems. The air cleaning second wave uses microfarms that take their carbon from the air or stored $CO_2$. Biosystems often use the bubbles from injected air and $CO_2$ to mix the culture.

SCAD also uses direct air capture, DAC as a carbon source. This mechanical- biosystem hybrid pulls large amounts of $CO_2$ directly from the air.

Ecolanda's extensive waste-to-energy systems are strong carbon suppliers to biosystem. The pyrolysis heat and pressure elements create flue gasses that are linked with biosystems for CCU. Several solid waste technologies use micro-organisms to generate syngas or green chemicals. Components of both outputs can feed carbon into microfarm cultivation systems.

### Mitigate nearby emissions

SCAD does not operate any coal or other fossil fuel power plants. Power plants and their nearby coal mines are terrible emitters of GHG and numerous toxic elements including deadly heavy metals. SCAD makes commitments to capture and recycle stack gasses and create remediation processes for areas destroyed by mines.

Microfarms can mitigate some emissions. Power plants run 24/7 while algae only grow and capture pollutants during daylight. Plants typically emit gasses faster than most biosystems can capture them. Therefore, biosystems offer only a partial solution to power plant emissions.

Many plants burn lignite, the lowest grade of coal. Lignite burns with low energy because the coal is poorly formed and mixed with considerable dirt.

### A smart energy bioproduct

Lignite and low-grade coal plants create three problems compared with higher-grade coal. They produce less heat and power even though they burn more material because the coal is laced with contaminants. They emit more GHG and carbon particulates from burning more material. In addition, the lignite is more expensive to mine.

SCAD tackles these challenges in reverse order. Higher mining cost occurs because lignite suffers a 30% loss in mining as the material turns to dust. These coal fines are not combustible and foul combustion chambers if not separated. The fines create a very expensive waste stream. Coal fines are easily carried on the wind and must be transported and buried before these minute coal particles cause deadly dust storms.

SCAD uses a coalgae solution developed at the Nelson Mandela Institute in South Africa.[35] Coal fines are fed into biosystems where algae coat the tiny particles. When dried and pelletized, the coalgae burn hotter than high-grade coal.

The natural algae coating adds oxygen to the furnace, creating more heat and energy. Higher heat burns cleanly and practically eliminates coal soot particulates. The coalgae trash-to-cash clean technology saves costs, decarbonizes the atmosphere and stops black soot pollution.

### Decarbonize everything

SCAD Ecolanda plans to decarbonize production across the board. The goal is not zero carbon, but carbon negative production. SCAD leadership believes it is not enough to prove carbon negative technologies. The value proposition and methods need to be clearly conveyed to world leaders.

SCAD targets the five energy sectors that emit the most carbon: energy production, food supply chain, transportation, buildings and industry. SCAD expands environmental preservation beyond the carbon footprint.

## *Positive Ecological Footprint*

A carbon footprint reports only the management or non-management of carbon emissions. The ecological footprint expands focus to forest, cropland and pasture preservation. It includes the ecological impact of the built environment, land with man-made structures. The eco-footprint also reflects concern about the health of fisheries.

SCAD Ecolanda designs production, construction, food, transportation, waste and energy systems to be carbon-neutral or carbon negative. Each sector benefits from smart technologies that create a Positive Ecological Footprint, PEF. Ecolanda agri-energy production comes from renewable sources. All-electric transportation moves on roads, rails and pathways covered with materials that create a PEF.

Building construction uses carbon-neutral and primarily biodegradable materials. Construction materials insulate buildings so effectively they reduce lifetime energy costs by 80%. At the end-of-life, which may be 50 years, the construction materials go through the BioRenew technology where their elements are repurposed into new construction materials or other bioproducts.

The smart ecocity uses exclusively PEF materials, energy and services.  The built environment benefits from links that eliminate emissions.

## *Climate care summary*

SCAD's climate strategies eliminate emissions on the Ecolanda campus. The Emerald flow reverses carbon flows from carbon discharge into the atmosphere to carbon capture in biosystems.

### Ecolanda Decarbonization

| ACTIVITY | METRIC TONS OF CO2e |
|---|---|
| MSW | 1.5 |
| LANDFILL | 1.5 |
| AGRIC. WASTE | 0.51 |
| DAIRY | 200MT/HEAD |
| MEAT | 100MT/HEAD |
| AQUACULTURE | 0.5MT/HEAD |
| BEEF | 100MT/HEAD |
| FOOD PROCESSING | 1MT/MT |
| VERTICAL FARMING | 100kg/CO2/sqMtr |
| GLASSHOUSE. | 1kg/CO2/sqMtr |
| BROAD-ACRE CROPS. | 100MT/1Hct |
| COAL FIRED P/P. | 10MT/MW |
| COALGAE | 2MT/MW |
| PV SOLAR. | 1MW=1000 CO2e/YR |
| THERMAL SOLAR | 1MW=1000 CO2e/YR |
| CSP SOLAR | 1MW=1000 CO2e/YR |
| VERTICAL WIND | 1MW=1000 CO2e/YR |
| DAC | 25MT |
| NRICH ALGAE | 2MT/MT |
| RACEWAY ALGAE | 2MT/MT |
| BIODIESEL | 3MT/ML |
| GREEN HYDROGEN | 10 |
| GREEN AMMONIA | 35 |
| ZERO CARBON HOUSING | TBD |

Ecolanda agri-energy-waste systems are designed with carbon capture and reuse in mind. SCAD decarbonization includes the following activities.

## *Carbon sources*

The decarbonization graphic summarizes the many carbon sources. CCU requires extensive planning, piping and recovery biosystems. Systems integration pervades all Ecolanda CCU biotechnologies. Smart sensors monitor and report inputs and outputs for carbon and other nutrient recovery, recycling and reuse.

A credible third-party analysis and validates ecological metrics. Industrial Economics, IEc in Cambridge Massachusetts receives live metric feeds. The IEc team analyses the big data sets to determine how much carbon gets captured and how reuse saves ecological damage.

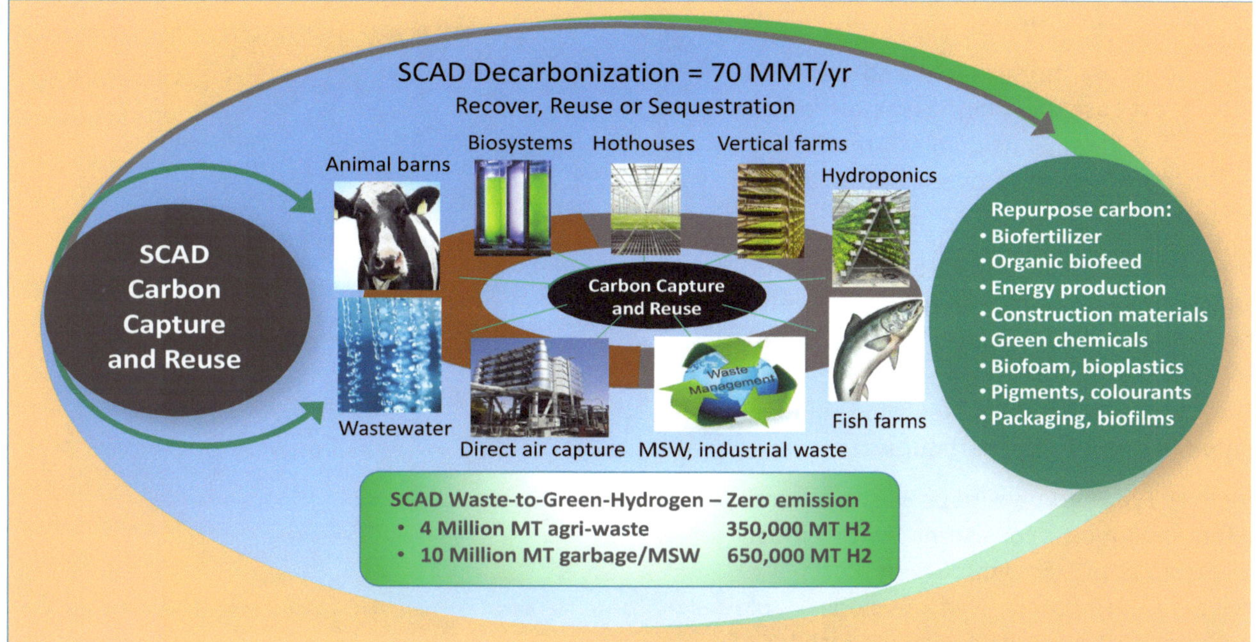


SCAD conscientiously practices ecological transparency. Ecosystem reports go to the host country government such as the Ministry of Interior or Environment and one or more academic institutions.

Ecolanda makes commitments beyond climate to care for all ecosystems. What could be stronger than avoiding carbon release and CCU?

## *Physical footprint*

Each Ecolanda campus varies in size based on land availability and national priorities. Several countries are planning a single campus of over 100,000 hectares. The total land footprint is close but not contiguous.

One country has plans for three sites with the hectares dedicated to each element shown in the table. The agri-energy facility in each is 3,000 hectares. The Nrich facility for microcrops consumes 500 hectares of non-crop land. The SEPS battery plant will fabricate and produce aluminium batteries.

SCAD plans to create a local smart agricultural college in a joint public-private partnership. The Education Centre and Innovation Park are also public-private partnerships.

**SCAD Operations Size for Three States** — SCAD

| State 100 Ha | 1. 8K | 2. 11K | 3. 12K |
|---|---|---|---|
| **Agri-Energy** | 3,000 | 3,000 | 3,000 |
| **Nrich Facility** | 500 | 500 | 500 |
| **Waste Treatment** | 1,000 | 1,000 | 1,000 |
| **Green Hydrogen** | 250 | 250 | 250 |
| **Green Ammonia** | 250 | 200 | 250 |
| **Smart Ecocity** | TBD | TBD | 950 |
| **Smart Community** | 500 | 450 | 600 |
| **Education Centre** | 50 | 50 | 50 |
| **Innovation Park** | 150 | 500 | 1,000 |
| **SEPS Battery Plant** | N/A | N/A | 1,000 |
| **Agricultural College** | 100 | 50 | N/A |
| **Energy Crops** | 2,200 | 5,000 | 4,000 |

### Create a clean BioHub

Most ports are noisy, dirty and polluted from heavy use by marine vessels of all types. The heavy mechanical machinery creates constant noise overlayed by traffic and fast-moving cranes.

Cargo and container ships are notorious for sending plumes of carbon into the atmosphere. Their engines are optimized for horsepower, not for clean air efficiency. Many ships leak diesel oil and human contaminants into the harbour waters.

The Port Types graphic maps waste, air and water pollution consistent with most industrial ports. Several Green Ports are under development with hopes of becoming carbon neutral in a few decades.

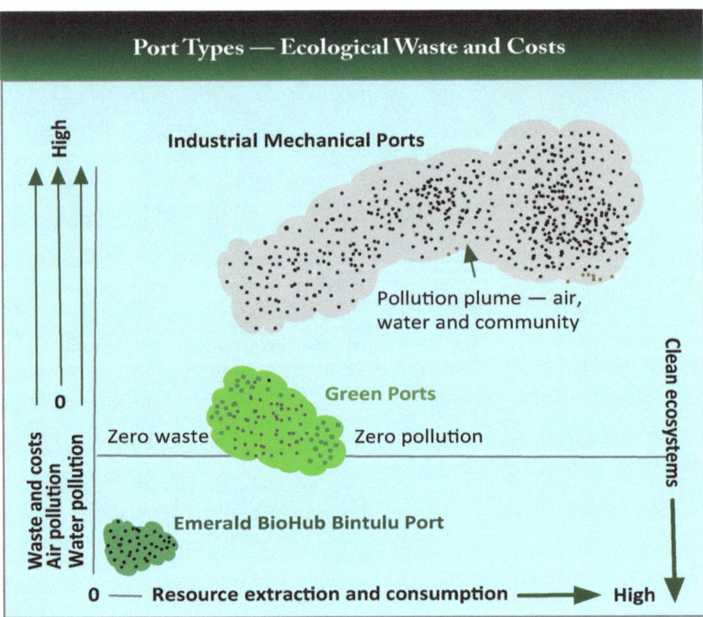

SCAD will build an Emerald BioHub port at Bintulu, Malaysia. The Emerald BioHub port will benefit from high efficiencies and a positive eco-footprint. The port will minimize natural resource extraction, which will occur primarily for constructing green energy systems. The BioHub will create zero waste and zero pollution.

The BioHub will be quiet, using mostly silent electric vehicles and engines. DAC and biosystems on site will captures and reuse millions of metric tons of $CO_2$.

Biosystems will clean and restore polluted water and local ecosystems. Harbour and coastal ecosystems sequester large amounts of carbon. A kelp forest can sequester up to 20 times more carbon per hectare than land forests. Marine plants that contribute to carbon sequestration.

Mangroves, kelps and seagrass live in rich soil. When these plants die, the leaves, branches, roots, and stems get buried underwater in the muck. The low underwater oxygen allows the plant material to stay buried for decades or longer. A kelp forest attract biodiversity that include filter feeders that also cleans the water.

# 12. Nrich Ecological Restoration

*Land degradation, biodiversity loss and climate change are three different faces of the same central challenge: the increasingly dangerous impact of our choices on the health of our natural environment. We cannot afford to tackle any one of these three threats in isolation – they each deserve the highest policy priority and must be addressed together.*

— **Sir Robert Watson**, Chair of IPBES[36]

**Ecological Restoration**

| Air | Water | Soil |
|---|---|---|
| CCU - two tons of $CO_2$-e for every ton of algae | Produce 20% more fresh water than is used | Bring dead, abandoned soil back to life |
| Smoke stacks - coal, cement, industry | Clear pollutants from water | Restore the full set of micronutrients |
| Black soot particulates | Remove pharmaceuticals | Restore soil porosity |
| Biocycle animal carbon | Remove heavy metals | Recover from salt invasion |
| Smog - nitric, sulfur and toxic x-oxides | Capture and clean agri-overspill | Restore eroded and worn out soil |
| Direct air capture - $CO_2$ | Eliminate water pollution | Restore soil biodiversity |

**N**rich enriches life by providing superior nutrition for people, animals, plants and ecosystems. SCAD Ecolanda goes further with a commitment to use their novel biological tools to restore degraded and destroyed ecosystems.

SCAD Ecolanda addresses biological ecosystem restoration with a six-layer, 3S3R strategy.

1. **Sanctuary** – create an Ecolanda Sanctuary that practices SCAD's six 3S3R strategies.
2. **Save** – ecosystems from exploitation and extraction of agricultural inputs.
3. **Substitute** – save ecosystems by substituting cultivation methods such as freedom foods and bioproducts that do not exploit cropland or verdant ecosystems.
4. **Recover** – emissions and pollution that would otherwise foul ecosystems.
5. **Rebuild** – fertility and soil structure in degraded ecosystems including macro and microorganisms.
6. **Restore** – flora and fauna to rebuild natural ecosystem biodiversity.

These six strategies work in harmony and drive Ecolanda design decisions. Biological ecosystem restoration does not happen by chance. Biorestoration requires advanced planning, conscientious design and rigorous systems integration.

## Ecolanda Sanctuary

SCAD will create a National Sanctuary dedicated to restoring biodiversity. The sanctuary, a public-private partnership, will benefit from, demonstrate and teach SCAD's 3S3R eco-restoration methods.

The Sanctuary will be located near or in the smart eco-city. Visitors will be able to walk trails on 10% of the Sanctuary. The majority will be closed to allow flora and fauna to flourish. Trails will allow robotic rovers to silently tour the Sanctuary. Visitors may use virtual reality headsets to see 360 degrees as if they were riding in the rover.

The Sanctuary will practice "rewilding." This strategy focuses on repairing biodiversity and ecosystem health by restoring natural processes. Rewilding lets nature take over and re-establish balance. The National Sanctuary will be named for a National ecological champion. The sanctuary will include an eco-learning centre where short courses will be available for people of all ages. The learning centre will include a biobank to preserve indigenous seeds and microorganisms.

A National Eco-Restoration Board of Directors will guide and lead the eco-restoration process. The Board will orchestrate return of precious flora and fauna threatened with extinction.

## Save

Ecolanda's ecosmart agri-energy production and freedom food saves ecosystems from the need for extraction of industrial agriculture inputs.

SCAD's Nrich biotechnology restores the atmosphere through biocycling carbon and other GHG which mitigates global warming. SCAD engineers off-takes from carbon emitters where the carbon flows to algae biosystems for capture and reuse. SCAD applies direct air capture of $CO_2$ in addition to Nrich biotechnologies.

Water serves the soul of ecosystems across the environment. Ecolanda operations save millions of liters of blue water, which benefits all stakeholders, including biodiversity.

SCAD's eco-friendly agri-energy methods save rural communities and ecosystems from the nasty plume of GHG and toxic agri-chemicals that invade their air, water and ecosystems.

## Substitute

The first step in industrial farming in the Spring is cultivation. Fields are cleared of nearly all life. The term "field cultivation" covers a multitude of sins. The brute force from huge machines rip deep into the soil, killing a majority of the crop-beneficial microorganisms. Cultivation intentionally leaves nothing living to compete with the crop.

Herbicides are applied to kill all competing seeds that may sprout after cultivation. Herbicides kill both seeds and microbial life. Chemical fertilisers then exterminate most the remaining beneficial soil microbes.

Cultivation occurs in the Spring and leaves fields highly vulnerable to erosion from annual winds and rain. Rain and irrigation leave soil wet. Heavy tractors, trucks and wagons repeatedly drive across the field compacting the soil. Years of repeated cropping leave soils devoid of beneficial microbes and severely compacted. Rain does not percolate down through compacted soil.

SCAD prefers no-till farming for broadacre crops to save beneficial microbes and to avoid soil compaction.

Growers use no or minimal herbicides. Nrich biofertilisers provide strong substitutes.

Freedom food cultivation in CEA saves ecosystems by simply not using cropland to produce food and other consumer products. Growing microcrops in vertical farms saves thousands of hectares from the insults inflicted by intensive mechanical agriculture.

SCAD avoids the use of inorganic fertilisers because they kill the symbiotic microorganisms in the soil and too often erode into local waterways.

Ironically, chemical fertilisers must be broken down by the microorganisms in the soil before the agri-chemicals can be assimilated by the crop.

Algae are one of the primary microorganisms that do the breakdown work in soils. Nrich skips the use of mined chemical fertiliser. Algae cells are pre-loaded with the full set of essential nutrients and are delivered to each plant in a form that is immediately bioavailable.

Substituting Nrich biofertiliser for inorganic chemicals saves costs and extensive ecological damage.

Water pollution from fertiliser overspill has created over 410 dead zones in coastal waters and lakes worldwide, affecting an area of 250,000 square kilometers, (95,000 miles$^2$. This is about the size of New Zealand. Nrich biofertiliser has no or minimal overspill, which saves waterways from eutrophication, loss of oxygen.

Herbicides, fungicides and insecticides used in agriculture cause pregnant mothers to abort or birth newborns with severe developmental disabilities. These toxic compounds cause both acute and often life-threatening poisoning and long-term chronic illness for consumers, farmers, farm animals and rural communities. Pesticide residuals cling to produce, which amplifies health risks for unsuspecting consumers.

Ecolanda uses Nrich biofertilisers that include natural biostimulants and biopesticides. Substituting natural biosolutions for agri-poisons saves millions of families from the tragedy of children that suffer developmental disabilities.

### Recover

Industrial methods have polluted the air with $CO_2$, black soot particulates, dust and toxic agri-chemicals. Nrich biosystems produce healthier food and bioproducts with zero emissions.

Ecolanda uses carbon negative biotechnologies that recover and repurpose rather than emitting carbon and other GHG.

Nrich cleans the hydrosphere through biocycling waste in polluted water and repurposing those elements into fresh, clean bioproducts. Algae biosystems can clean nearly any water source – domestic, industrial, medical and even mining.

Mines and deep wells may use advanced Nrich biotechnology to capture and remove poisonous heavy metals before allowing the biosystem to do its biocycle work. Biosystems recover and recycle wastewater nutrients and toxic compounds that would otherwise pollute ecosystems.

### Rebuild

Industrial agriculture makes a classical mistake by intensively farming cropland for decades. Healthy soil contains trillions of living microorganisms. Continuous cultivation and compression from heavy equipment paired with nutrient extraction by crops leaves the soil depleted, compacted and exhausted.

Farmers typically replace only the top three NPK fertilisers – nitrogen, phosphorus and potassium. The other 21 essential macro and micronutrients are ignored. Each crop consumes about half the NPK nutrients, which are lost to the field with the harvest. Crops continually deplete micronutrients and the humus that holds soil moisture. Nearly all of the NPK residual erodes on wind and water or drop below the root zone where they are no benefit for crops.

Next season, the farmer must buy and apply the NPK fertiliser again with a new round of cultivation. Currently, industrial agriculture has no practical means to rebuild either the micronutrients or the humus.

Nrich biofertiliser rebuilds the soil fertility and structure from the foundation upwards. The subsurface drip system delivers the full set of vital macro and micro-nutrients. Crops respond with higher yields and higher quality produce.

Industrial agriculture systemically compresses soil until it will no longer support crop roots. Nrich can improve soil porosity. Nrich biofertiliser improves soil structure, which allows for stronger and longer root development. Many of the high-nutralence algae cells enter the plants and improve the crop. Indigenous local algae cells continue to grow in the soil creating rich organic matter, humus and plant sugars. Plant sugars act like honey and attract trillions of additional microorganisms, which create natural symbiotic communities in the looser soil.

Abundance growers improve their fields every year with Nrich biofertiliser. Nrich has the unique capacity to restore soil fertility and even bring dead soil back to life. Growers may address each of the soil maladies shown.

Nrich reconstructs soil quality and structure by increasing nutrient levels and humus in one or two years. Nrich flips compaction to looseness.

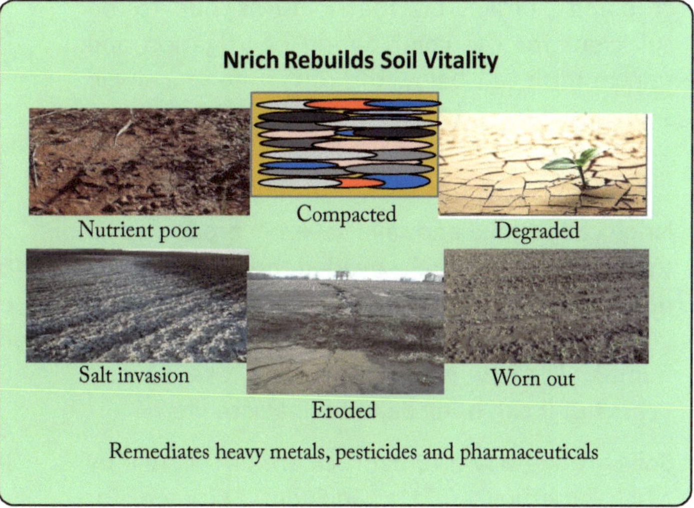



Nrich biofertiliser delivers 74 nutrients to the soil. Indigenous algae that delivered the nutrients in algae cells continue to grow in the soil, creating rich organic biomass, humus.

Algae biomass attracts many other micro-organisms. The highly diverse micro-community works symbiotically to improve soil structure that benefits fertility and crop growth.

Nrich biofertiliser allows farmers to leave every field better than they found it after every harvest. Algae-based nutrients and humus continue to build layer upon layer of fertile soil with every crop.

## Salt invasion

Hundreds of civilizations of all sizes have crashed from a salt invasion.[37] Irrigation water flows long distances and picks up salt. Dissolved salts crystalize in irrigated cropland. When too much salt remains on or in the soil, crops die, children die and then the civilization crashes.

The only historical remedy for soil salt invasion has been to pray for a farmers' rain — long and slow — to flush out the salt. Unfortunately, global climate chaos creates more severe storms that flood rather than soak out salt.

Field trails have demonstrated that Nrich improves porosity in compacted and salty cropland 500%. This metric turns out to be sufficient for the combination of irrigation and rain to flush salt below the root zone in a single season; where it does no crop damage.

## Limitation

Nrich can rebuild and save severely eroded soil if the next level below the eroded topsoil is dirt and not rock. Subsurface drip systems can deliver algae biofertiliser and restore nutrients and humus. Fertility restoration may take several years but it can bring dead soil back to life.

Soils eroded down to bedrock are **not fixable** by Nrich methods. This includes surface or mountaintop mines that have scraped off the topsoil.

SCAD offers a solution nature uses for the world's hottest and driest deserts. Nature lays down a thin layer called an algae crust that holds the topsoil layer firmly and resists erosion.[38]

## Restore

SCAD Ecolanda restores health to ecosystems and biodiversity. Nrich restores health to degraded ecosystems, which allows the recovery of biodiversity. Healthier nutrition and thriving ecosystems attract extensive biodiversity.

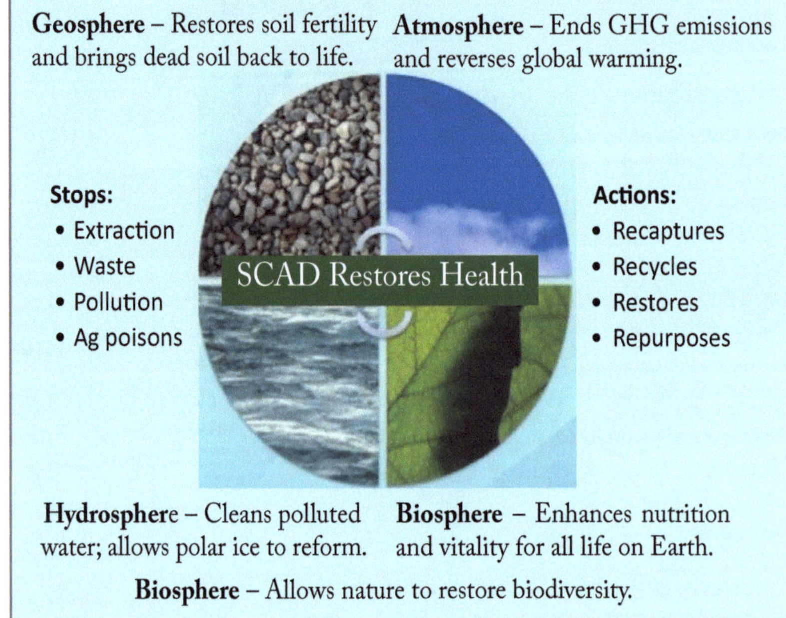


Nature may require a century to create a centimeter of topsoil. Nrich methods may restore fertility in one or two growing seasons. Once fertility is restored, abundance methods ensure continuous soil improvements year over year.

SCAD makes extensive use of sensors to monitor soil fitness for purpose. Field and CEA experience allows growers to create nutrient recipes optimized for specific plants in precise growing conditions. Advanced analytics are applied to big data sets in order to optimize fertility.

Metrics and analytics allow SCAD to first restore ecosystems and then find ways to continually improve environmental health.

Industrial farmers constantly degrade their fields and neighboring environments. Abundance farmers continually enhance the health of their soil and their community's ecosystems.

The next section proposes an initiative to end deforestation and repopulate our forests and their natural colorful biodiversity.

# 13. Emerald Forest Initiative

*God has cared for these trees, saved them
from drought, disease, avalanches and a
thousand tempests and floods. But he
cannot save them from fools.*

**– John Muir**

Farmers are neither fools nor foolish. Farmers are driven to remove trees by their need for feed – food for their families. Only three commodities are responsible for 90% of deforestation: soybeans, palm oil and beef.

What if farmers were able to make a better living growing microcrops on non-crop land? Would compelling economics eliminate the need to clear forests?

SCAD's commitment to ecology includes the goal of the global **Emerald Forest Initiative**, dedicated to recovering our forests.

Current strategies have failed forests. Eco-shaming does not work. Deforestation continues to increase rather than decrease. The *Guardian* found that our planet loses a football field of forest every six seconds 24 hours a day.[39]

SCAD Ecolanda employs a four-layer strategy to stop deforestation.

1. **Demonstrate and train** people in abundance methods that require neither natural resource extraction nor cropland.
2. **Educate, mentor and support farmers** and their families to learn and use eco-friendly abundance methods.
3. **Sponsor Emerald Forests**, an international competition calling for innovators to prove how microcrops can provide farmers with more income and less risk without abusing forests or cropland.
4. **Invent a green alternative to fuelwood** for heating and cooking. SCAD will teach people how to use microfarms to make cooking and heating oil that burns cleanly.

## Deforestation

About half of the world's tropical forests have been cleared since 1960. All their stored carbon has been released to the atmosphere. An area the size of Switzerland (38,300 square km or 14,800 square miles) is lost every year.[40]

Deforestation adds 30% to all GHG emissions.[41] These add to agriculture's 26% contribution to total GHG emissions.[42] Forests contribute nearly 30% of atmospheric oxygen while 70% comes from algae, largely phytoplankton in the oceans.[43]

Deforestation disrupts the water cycle, resulting in changes in precipitation and erosion into waterways. Trees extract groundwater through their roots and release it to the atmosphere.

When a forest is removed, the trees no longer transpire water. The atmosphere becomes drier and hotter. Deforestation reduces water stored in the soil and severely depletes groundwater. The result decreases moisture in the atmosphere and makes ecosystems hotter and drier.

Scientists found that in China's deforested north precipitation levels decreased significantly.[44] From the 1950s to the 1980s, precipitation levels decreased by 33%, making the area substantially hotter and drier.

Nearly 75% of the Earth's freshwater comes from forested watersheds. Forest loss degrades water quality and availability. A 2020 UN report on the world's forests found that over half the global population relies on forested watersheds for their drinking water.[45]

Those waters are under siege. Forests need a minimum of 60% cover to hold the soil and prevent landslides.[46]

### Forest land replacements

Replacement crops such as soybeans and palm oil have weak root systems and cannot hold the soil. Row crops accelerate erosion and moisture loss.

In some hilly countries, such as Haiti, over 75% of cropland that replaced forests has been abandoned due to erosion.[47] Slash and burn farmers that clear-cut a hillside may find the land holds fertility for only one or two seasons. Erosion forces farmers to move on and clear more forest, which continues the cycle of soil loss.

### Biodiversity loss

Tropical rainforests are the most diverse ecosystems on earth. Rainforests harbor more than 80% of the world's known biodiversity. Deforestation causes the loss of 137 plant, animal and insect species every single day, about 50,000 species a year.[48]

Farmers growing crops on deforested land know they are damaging their children's legacy. Farmers would like to know a better way to farm and to make a living.

SCAD plans a novel strategy to end deforestation that aligns with Ecolanda's vision and values.

### Ecolanda Emerald Forest ePrize

SCAD will sponsor the **Ecolanda Emerald Forest ePrize** designed to save and restore forests while ending deforestation. The Emerald Forest ePrize will convey a simple message:

*The day forest-land farmers can produce superior microcrop substitutes at lower cost than field grains, deforestation ends.*

Replacements include better oil than palm, better animal feed than soybeans and plant-based meat.

The Emerald Forest ePrize will award $100,000 to winners. The SCAD Ecolanda website will provide details, including production metrics for palm oil, soybeans and beef. SCAD will post relevant eco-footprint and economic metrics that enable clear comparisons.

Four Ecolanda Emerald Forest ePrize categories are designed to save forests. Entrants may use narratives, graphics, pictures or videos.

1. **Create a children's booklet** or cartoon that shows how ***Anna, the Tiny Mighty Algae***, saves our forests by producing substitutes for soybeans, palm oil or beef.

   Two examples available free here, show how Anna has already saved the world twice.[49] She made oxygen by eating carbon dioxide. She became the first rung on the food chain. Today algae are eaten by 1,000 times more consumers than any other food because algae delivers superior nutrition. Entrants may use this Anna character, only for this competition, or create their own.

2. Share a description of your **microfarm design**. How does the biosystem compare on its eco-footprint with soybeans or palm oil? (Ten ecofootprint and five economic criteria are provided.)

   SCAD will provide standard models that allow growers to plug in their metrics.

3. Share a description of your **plant-based meat**. How does your process compare on eco-footprint, economics, taste, texture and other dimensions with animal meat?

4. **Create architectural designs** for imaginative future sustainable food cultivation systems. Designs and descriptions may be public art, urban gardens, vertical farms or other ideas.

## *Emerald Forest Prizes*

Grand prizes will be awarded in each category. The top ten entries in each category will be published in order to convey the urgency and methods designed to end deforestation.

SCAD executives Robert Henrikson and Mark Edwards orchestrated a successful international algae competition. Winners and great graphics are available at AlgaeCompetition.com. Graphic artists and architects invested over 20,000 hours creating images. This project resulted in a book with beautiful visionary architecture, descriptions of biosystems and fabulous algae-based foods.

SCAD expects the Ecolanda ePrize will marshal public sentiment to stop deforestation.

## *Other SCAD initiatives*

SCAD's support for Emerald Forests goes beyond the ePrize. The competition serves as a form of marketing research to find out how people are attacking deforestation and the progress they have made. SCAD will share this valuable market intelligence globally.

SCAD Ecolanda will provide in-person and distance learning for farmers globally to learn abundance methods and how to grow microcrops that save forests.

SCAD scientists will assist with training and tools that support ecological restoration and rebuilding forests. A major effort will focus on how to restore biodiversity without harm to sister creatures.

SCAD plans to implement a series of aligned strategies in the future. The Emerald Small Fish Initiative will focus on saving the small fish in the sea by promoting eco-friendly, algae-based Omega-3 oil to replace Omegas sourced from fish.

Another Ecolanda climate change strategy will address high GHG emitting materials such as cement, asphalt, bioplastics and construction materials. SCAD plans to replace these eco-dirty products with clean biodegradable substitutes.

Ecolanda climate initiatives will demonstrate to the world not just that microcrops offer strong climate solutions but also abundance methods are safer and more profitable for growers. Abundance methods are healthier for growers, their communities and the environment.

Designs below are from the Algae Competition.

## Forest product substitutes

Industrialized countries consume 12 times more wood and wood products per person than the non-industrialized countries. Almost half of the world's timber and 70% of the paper is consumed by Europe, the United States and Japan.

SCAD Ecolanda plans to cultivate superior substitutes for wood and paper products. The paper industry is fourth largest in producing GHG and among the leading industries degrading water. Paper products contribute substantially to deforestation and the related economic, social and environmental costs.[50]

SCAD will use the considerable fibres in algae to cultivate paper products that provide better solutions for reading, writing, wrapping and packaging.

Packaging production consumes huge amounts of energy, water and other natural resources. Packaging creates billions of tons of waste materials that pollutes our air, water, soil and oceans.

Algae-based substitutes – cardboard, biofilms, paper and bioplastics – serve as superior biodegradable replacements with zero emissions or pollution.

Buildings consume massive amounts of timber which diminishes forest stocks. SCAD will invent several biodegradable building materials with superior strength, usability and construction attributes to wood.

Algae construction-grade biofoam offers three times the tensile strength of wood. In addition, biofoam allows buildings to be constructed in less than half the time and provides 80% better insulation from noise, bugs and weather.

Algae-based construction material substitutes currently are not cost competitive. SCAD will improve productivity, which will change the economics. When the economics flip, biodegradable packaging and construction materials made with abundance methods will be dominate construction markets.

## Fuelwood

People in sub-Saharan African countries cut and burn firewood several times faster than tree growth rates. Families must constantly travel further into the forest to find wood. Some families spend most their day searching for firewood.

Harvesting firewood indiscriminately expands deforestation and reduces timber resources. Timber reduction creates loss of habitat and leads to extinction of the species that once lived in the degraded habitat.

Half the people on earth tonight will cook their supper over an open fire. Most cook over an open fire of wood, animal dung or coal in a small enclosed kitchen.

Household air pollution causes stroke, heart disease, asthma, chronic obstructive pulmonary disease and lung cancer.[51] Over 5 million people die prematurely every year from black smoke pollution due to smoky cooking and heating stoves. About half the smoke deaths are caused by pneumonia among children under five. Millions more children and their grandparents are disabled by black soot particulates inhaled from household air pollution.

SCAD will cultivate clean burning cooking and heating stove oil. Algae oil provides higher heat per kilogram than wood. Algae oil is easy to use and burns cleanly, with no black smoke particulates. Algae oil does give off a light odor similar to French fries. 

SCAD Ecolanda will train farmers and families how to use abundance methods in microfarms to grow cooking and heating stove oil, biofertiliser and animal feed. Local training will provide independence and health to many in Ecolanda's neighboring communities.

SCAD's Emerald Forest initiative will invent natural microcrop substitutes for forest products. This will allow regions to reforest and their colorful biodiversity to re-establish and flourish.

# 14. Emerald Renaissance

*The Emerald Renaissance provides a clean, renewable alternative to fossil resource-based industrial mechanical production.*

The industrial revolution produced cheap food and consumer goods. The extremely high ecological cost came from extracting and consuming massive amounts of natural resources and then discarding wastes into the environment. Industrial production imposes an exceptionally high cost on social and economic development and destroys the environment.

The Emerald Renaissance counters the industrial mechanical production with biological cultivation. Biosystems avoid natural resource extraction and consumption and clean rather than foul the environment.

Ecolanda growers are able to produce a wealth of bioproducts using biosystems instead of mechanical production. Growers can cultivate microcrops to produce nearly anything made by industrial means.

## Why?

The first question many people ask is:

*If biological systems are superior to industrial mechanical production, why do producers continue to use industrial methods?*

Industrial producers find industrial products are cheaper to create than bioproducts. The industrial cost advantage occurs as an artifact of current social policy. Society places no tax on extraction. Industrial producers are able mine resources and the dump or burn their wastes indiscriminately with no tax or cost.

Ironically, instead of paying the full social cost for industrial production, society subsidizes inputs that reduce production costs but increase waste. Industrial producers pay only a small fraction of the social cost for their energy, fuel and water. They pay zero for their waste streams.

Future generations will pay those taxes in the form of much higher costs for scarce resources. Our children and their children will pay extraordinary costs for the impacts of human-caused climate change and polluted air, water and ecosystems.

If society charged industrial producers for fossil resource loss and environmental damage, microcrop bioproducts would be 30% less expensive than industrial products today.

## Ecolanda bioproducts

Ecolanda will demonstrate to the world the benefits of biosolutions. Growers will cultivate microcrops for food, feed, fertiliser, green chemicals and energy products. Cultivation will include health products, pharmaceuticals, nutraceuticals, cosmeceuticals and medicines.

Along with bioproduct cultivation, growers will decarbonize the atmosphere, clean water and remediate degraded and destroyed ecosystems.

## Food

Ecolanda growers will produce a wealth of conventional food, enough to feed five million consumers. Growers will cultivate microcrops such as algae-based foods, food ingredients and functional foods that deliver superior nutrition and health benefits. These foods increase resistance to disease vectors and fight disease when they attack the body. Consumers want to eat healthy whole foods close to the food's natural state.

Algae food ingredients have supplied stronger color, flavor and texture to foods for millennia. Most ingredients today come from sea vegetables, macroalgae, because they have the longest history in foods.

Macroalgae, seaweeds, are highly visible and accessible. Our ancient ancestors commonly ate sea vegetables because they were nutritious and easily gathered along seashores and lakes.

Algae components are intensely integrated in modern food supply chains. An Arizona State University study performed a market basket test across seven grocery stores.[52] The research found that 72% of processed foods included one or more algae compounds.

Algae ingredients in modern food include:

**Beer and diet sodas** as a clarifier to remove haze-causing proteins. Frozen foods – pies, pastries fillings, yogurt and ice cream as an emulsifier.

Dairy – whipped toppings, milkshakes, skim milk, evaporated milk, chocolate milk, ice cream, cheeses, cottage cheese, infant formulas, custards and instant breakfasts. High protein drinks – healthier protein, vitamins, minerals and micronutrients.

Fruits – fruit juices, syrups, jams and jellies. Sauces, gravies and soy milk – thickeners and emulsifiers. **Pâtés** and processed **meat** — substitute animal fat with low calorie alternatives.

### *Algae vitamins*

Algae absorb a wealth of mineral elements that concentrate about one third of its dry biomass.

The mineral macronutrients include sodium, calcium, magnesium, sulfur, potassium, chlorine, and phosphorus. Micronutrients include iodine, iron, zinc, boron, copper, selenium, molybdenum, fluoride, manganese and nickel. The biomass also includes high-quality vitamins, other minerals and trace elements.

One tablespoon of algae powder provides the same amount of calcium as ½ cup of milk, 1½-cups of soybeans, 11 carrots, or 22 tomatoes. A tablespoon provides the same amount of magnesium as 2½ cups of milk, ½-cup of soybeans 9 carrots or 5 tomatoes. A tablespoon gives the same amount of iron as 32 cups of milk, ⅓ cup of soybeans, 11 carrots, or 5 tomatoes.[53]

Algae provide a mineral profile superior to that of land plants, milk, eggs or soybeans.[54]

Mineral availability from land plants, particularly legumes and grains, becomes compromised by phytic acid. Phytic acid binds the minerals, thus rendering them unavailable for absorption into the blood stream. In one investigation, phytic acid was undetectable in four species of algae. Iron absorption was 3.5 times higher for algae compared to rice.[55]

**Algae microcrop nutrient levels vs popular foods per kg**

| Nutrient | Microcrops have _ times higher than |
|---|---|
| **Proteins** - building blocks for bones, muscles, cartilage, skin, and blood. They help manufacture enzymes, hormones, and vitamins. | 2x > than soy<br>3x > than beef, fish, pork<br>6x > than eggs |
| **Iron** carries oxygen in the blood. Many women in childbearing years have iron-deficiency anemia. | 30x > than beef, fish, pork<br>65x > than spinach |
| **B12 vitamin** helps the body release energy, regulates the nervous system, aids in the formation of red blood cells, and help build tissues. | 3 to 4x > than animal liver |
| **Magnesium** heart rhythm, build bones, maintains the immune system and normalizes blood pressure. | 2x > than spinach<br>5x > than tomatoes |
| **Calcium** regulates nerve transmission, blood clotting, hormone secretion and muscle contraction. | 10x > than milk |
| **β-carotene**, pro-vitamin A, boosts immune system, helps skin, eyes, protects against CHD and cancer. | 5x > than carrots<br>40x > than spinach |
| **Chlorophyll** helps fight cancer, speeds wound healing, cleanses liver of toxins, improves skin, digestion and weight control. | 30x > than spinach<br>20x > than wheatgrass |
| **EPA and DHA,** omega-3 fatty acids improve eyesight, brain function and protect from CHD. | 1,000x > than any land plant<br>1,000x > than beef, poultry |

Algae iron is easily absorbed by the human body because its blue pigment, phycocyanin, forms soluble complexes with iron and other minerals during digestion. Phycocyanin makes algae iron more bioavailable. Unlike iron derived from terrestrial plants, the bioavailability of algae iron is comparable to that of heme iron in meats.[56]

Functional foods beneficially target functions in the body, beyond adequate nutritional effects. Common functional food enrichments are long chain omega-3 fatty acids-DHA/EPA, flavones, beta-carotene, lutein lycopene, fibre, proteins, catechins, beta-glucan and anthocyanins. Functional foods take all the forms of conventional foods including breads, soups, stews, pasta, beverages and snacks.

Major food retailers such as Amazon and Walmart are building supply chains that deliver food, medicines, pharmaceuticals and cosmeceuticals. Consumers will have the freedom to choose foods that prevent illnesses and other foods that treat specific diseases.

Market research shows that people strongly prefer following the advice of Hippocrates:

> *Let food be thy medicine and*
> *medicine be thy food.*

Algae support the global food system as useful ingredients and valuable compounds for functional foods. Consumers will choose algae foods for their extensive medical benefits as well as freedom from allergens and empty calories.

## *Taste*

Taste represents the critical variable for new food adoption. Consumers adopt functions ONLY if the food meets taste expectation. Algae provide of all the essential nutrients, vitamins, minerals, and trace elements essential for health and vitality. Algae can satisfy any appetite with a broad spectrum of aromas, colors, tastes, and textures.

Algae's high glutamic acid content stimulate taste receptors, amplifying taste differentiation. This increases the desire to consume algae foods for good taste. Most functional food studies develop new practical uses from algae nutrient extracts that enhance food palatability. Kelp is added to beer to enhance the malty taste while other algae components improve liquid clarity.

## *Plant-based meats*

Ecolanda will create plant-based meats, PBM, for every taste. Each bite of PBM delivers 2.5 times more protein than beef. Unlike beef, these meats supply healthy oils, omega-3 EPA/DHA. They contain less fat, lower calories and great taste.

PBM reduces health risks for obesity, diabetes, cancer and heart disease. They are healthier than tofu or texturized veggie protein which are often used in synthetic meats today.

These foods allow omnivores to diversify their diets. They contain no allergens, contaminants or pesticide residuals.

PBM reduces consumer prices. They reduce social costs with zero consumption of water, cropland, fossil fuels, inorganic fertilisers or agri-poisons.

Global livestock emit 7.1 gigatonnes of $CO_2e$ per year. Livestock are responsible for 14.5% of all anthropogenic GHG emissions.[57] One cow-calf pair emit more GHG annually than a standard car traveling 20,000 kilometers.

PBM cultivation decarbonize the atmosphere, clean water and provide beautiful color for animals like flamingos.

Several algae species, like *Haematococcus pluvialis* react to environmental stress by metabolizing valuable oils, such as astaxanthin. The cells spontaneously produce the oil to protect itself from too much sunlight. Astaxanthin can reduce free radicals and oxidative stress and help the human body maintain a healthy state.

Astaxanthin provides clean healthy oils to animal feed and gives the beautiful pinkish-red color to salmon, shellfish and flamingos.

### 3D printed food

Tiny algae fines make a perfect food construction material for 3D printers. These devices offer fast,  consistent and flexible formation of PBM and dairy products. They also make reliable and delicious algae-based pasta, pizza and pastries.

### Human and animal health

Algae foods and ingredients optimize health. The proof lies in the addition of target compounds to functional and health foods. Many health promoting algae compounds are not available in conventional fossil foods grown in soil.

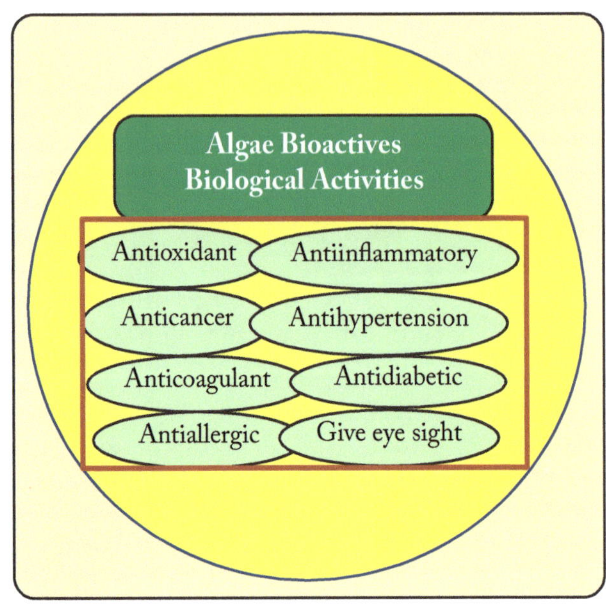


| Algae Compounds Provide Value for Functional Foods | | | |
|---|---|---|---|
| **Deficiencies** | **Major Organs** | **Major Systems** | **Diseases** |
| Micronutrients | Brain | Cardiovascular | Blood pressure |
| Vitamins | Eyes | Digestive | Hyperlipdemia |
| Minerals | Blood | Endocrine | Bleeding gums |
| Trace elements | Heart | Immune | Infections |
| Antioxidants | Lungs | Respiratory | Inflammation |
| Hormones | Kidneys | Circulatory | Cancers |
| Dissorders | Skin, hair, nails | Urinary | Immune |
| Mood | Hypothalamus | Nervous | Viral infection |
| Anxiety | Pancreas | Muscular | Bacteria infection |
| Psychotic | Liver | Integumentary | Injuries, burns |
| Personality | Thyroid | Reproductive | Diarrhea |
| Sexual | Pituitary | Skeletal | Diabetes |
| Development | Nerves | Lymphatic | Obesity |

Field studies show that a 3-week treatment with a tablespoon of algae a day brings nutrient concentrations to healthy levels.

A cluster of conditions causes metabolic syndrome — increased blood pressure, high blood sugar, excess body fat around the waist and abnormal cholesterol or triglyceride levels. These tend to occur together, increasing risk of the common diseases. The incidence of the metabolic syndrome steadily increases worldwide. Algae-foods repair metabolic syndrome.

Several Asian countries where sea vegetables are commonly consumed exhibit very low metabolic syndrome rates. Several studies show that diets that include only 6 grams per day of algae improve critical inflammation biomarkers and blood lipids.[58]

### Medicines

Micronutrients may be the most important compounds found in algae biomass. In many countries, including the US, over 85% of people suffer from micronutrient deficiencies.

Sodium alginate and other sea vegetable compounds are used in culinary physics at some of the best restaurants in the world.[59] Research shows sodium alginate pulls heavy metals including radioactive toxins from the body, such as iodine-131 and strontium-90.[60]

## Bio-construction materials

Construction materials account over 8% of global GHG and has a much higher percentage of buried or burned wastes and hazardous materials.

Cement mining, mixing and applying emits about 4% of global GHG. Algae-based cement can stop those emissions. The carbon is sequestered as long as the material stays in place - several generations. At end-of-life, deconstructing cement structures, GHG emissions and wastes add substantial ecological costs to cement.

**Asphalt** may create more GHG than cement but is currently not included in climate models. The two-part construction process begins with the production of aggregates, asphalt and cement. The second step includes mixing, transportation, paving, compacting and curing the cement stabilized aggregate. At the end-of-life, most asphalt is deconstructed and buried.

Algae-based asphalt provides a green substitute for conventional petroleum asphalt. At the end-of-life for cement or asphalt, algae biomaterials can go through the BioRenew process to create new biomaterials or other bioproducts.

**Construction-grade biofoam** made from algae oil delivers 2x the tensile strength of concrete block and 3x the strength of wood construction. Biofoam is highly insulating and delivers several proofs: sound, odors, bugs, fire, earthquakes and wind. Insulation delivers an 80% lower cost of lifetime ownership saving utility costs. Biofoam cuts construction time by more than 50% and does not require skilled construction tradesmen.

**Resin composites** may replace steel beams with superior strength. Advanced biomaterials can replace conventional industrial materials.

Packaging creates massive waste. Biopackaging is strong, resilient and entirely recyclable with BioRenew. Bioplastics may be used for biodegradable packaging, windows, construction or many other applications. Bioedible packaging, bowls, chop sticks or knives, forks and spoons may be eaten or biocycled after use.

## Bio-textiles

Textile production is the world's second most polluting industry after the oil industry. Total GHG emissions from textile production, 1.2 billion tonnes annually, represents 10% of global GHG. Textiles emit more than all international flights and maritime shipping combined.[61]

Textile mills generate one-fifth of the world's industrial water pollution.[62] Mills use 20,000 chemicals to make clothes, many of them carcinogenic. These toxic wastes typically run untreated into local waterways, damaging local residents' health. Textile factories produce over 5 billion tons of soot annually by burning coal.[63]

The primary raw material in the textile industry, cotton, consumes 20,000 liters of water to produce a cotton T-shirt and jeans. Cotton production uses 3% of annual global water consumption.[64] Industrial cotton production requires massive use of fertilisers, which contaminate local water bodies and cause dead zones. More pesticides are used in industrial cotton production than in any other crop.

Pesticides impose severe detrimental effects on young children. Children's immature livers and excretory systems may be unable to effectively remove pesticide metabolites. Research shows that low levels of pesticide exposure can affect young children's neurological and behavioral development. Evidence shows a link between pesticides and neonatal reflexes, psychomotor and mental development and attention-deficit hyperactivity disorder. Children exposed to pesticides suffer elevated rates of leukemia, brain cancer and soft tissue sarcoma.[65]

Algae contain plentiful fibres which can substitute for cotton, flax, bamboo and other textile sources. Algae textiles release zero GHG emission, clean wastewater during cultivation and use no pesticides.

Algae-based finishes and dyes for the textile industry present clean, non-allergenic, renewable and non-polluting alternatives to industrial methods. Algae dyes are typically brighter than industrial dyes.

### *Support a trip to Mars*

Ecolanda will have a Learning Centre focused on how microcrops benefit society. Guests will have an opportunity to see how microcrops can support a trip to Mars and deep space.

The Ecolanda Learning Centre will apply the recommendations from the NASA 100-Year Starship project.[66] Algae and her sister microcrops can provide food, life-support, waste cycling, green chemicals, bioenergy, and building materials.

A 100-year Starship cannot possibly carry enough food for a trip to Mars or beyond. NASA calculates that six liters of algae water can produce 600 grams of food or 2500 calories. This is the average daily food requirement for an adult male. The six liters of algae water can simultaneously produce 600 liters of oxygen and consume 720 liters of $CO_2$.[67] The carbon from the waste $CO_2$ gets transformed into hydrocarbons in food.

Human and animal waste products will need to be biocycled during the entire 100-year Starship mission. Gas waste, mostly $CO_2$ and methane, $CH_4$, are biocycled into food. Liquid wastes are biocycled into a host of clean bioproducts while algae clean the water for reuse. Human and animal solid wastes are biocycled into bioproducts for Starship life support.

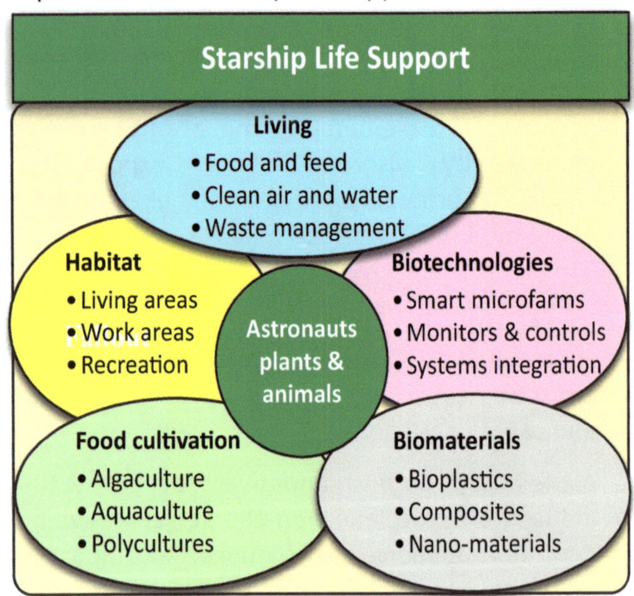


The Ecolanda Learning Centre will highlight how microcrops can provide excellent medical support for the Starship crew. The crew are in top physical shape and want to maximize their health and vitality.

Algae-based foods can be made into nearly any form, including ice cream, chocolate cake and plant-based steaks that imitate nearly any type of meat. Steaks are made with 3D printers that weave algae fibres into the meat to match the desired texture.

Algae-based functional foods, nutraceuticals, cosmeceuticals, pharmaceuticals and medicines sustain astronaut health. Functional foods help astronauts avoid illness while medicines fight disease threats when they penetrate the body.

Crew members will have an ongoing supply of recommended probiotics, antioxidants and phytochemicals.

The Starship cannot possibly carry a pharmacy will all the necessary medicines. A novel Intelligent Algae Repository, (IAR) maintains a vast algae library of 600,000 algae cultivars that can be searched for novel compounds. Each cultivar contains about 1 million cells of the algae species that fit in 1 $mm^2$. Algae cultivation can grow the desired compounds quickly from the sample. The target compound can be extracted to resolve the malady. The IAR can produce a nearly infinite set of medicines just in time when they are needed.

The next section examines Ecolanda benefits.

# 15. Ecolanda Benefits

*Simplicity is the ultimate sophistication.*
*— Leonardo da Vinci*

**B**enefits derive from Ecolanda's primary goal: *Lift the standard of living for everyone while improving the environment.*

Ecolanda listens and learns from each country's priorities for human and environmental health, economics and social. Project plans align with the country's goals to maximize benefits.

## Stewardshlp

Human and environmental health are closely integrated. Polluted ecosystems, cities and towns impose a severe health toll on citizens.

SCAD makes commitments to remediate and restore healthy environments. The pace at which Nrich and BioRenew clean ecosystems can be tracked by National leaders. After 18 months, enough data exists to support National leaders to set new goals for carbon neutrality and clean environments.

Ecolanda ecological data will support other National goals such as water pollution and use of pollution sources such as chemical fertilizers and pesticides.

## Blue water

Half of the world's hospital beds are filled with people suffering from a water-related disease.[68] Millions lack adequate access to blue drinking water. In developing countries, about 80% of illnesses are linked to poor water and sanitation conditions.[69] Healthy living requires safe, clean and affordable water.

Ecolanda not only saves water in agri-energy production but also creates extra blue water. The extra blue water will be shared with those most in need. SCAD will educate farmers in abundance growing methods that use Nrich biofertilizer that does not create toxic pollution in waterways. Abundance methods also minimize the need for pesticides and agri-poisons.

## Malnutrition

One of every two people on earth, 3.8 billion, are food insecure, without reliable access to good food.[70] Over 9 million people die of hunger each year; more than the death toll for malaria, HIV/AIDs and tuberculosis combined.[71] Hungry people consume fewer than 2,000 calories a day.

Some communities must endure the severe pain and suffering associated with endemic malnutrition. Global hunger figures understate the much larger number of people who are malnourished due to the lack of quality food or suffer from micronutrient deficiencies.

Clinical signs and symptoms of micronutrient deficiencies include:[72]

- **Iron** – Fatigue, anemia, irregular heartbeat, dizziness and headaches.
- **Iodine** - Goiter, developmental delay, and mental retardation.
- **Vitamin D** - Poor growth, rickets, and hypocalcemia.
- **Vitamin A** - Night blindness, dry eyes, blindness and poor growth.
- **Folate** - Glossitis, anemia (megaloblastic), and neural tube defects.
- **Zinc** - Anemia, dwarfism, hepato-splenomegaly and hypogonadism.[73]

The high prevalence of hunger among women has led to malnutrition becoming the leading cause of death for children. About 3.1 million children die from hunger each year.[74] Poor nutrition accounts for roughly half the deaths for children under five.

SCAD Ecolanda plans to partner with National medical leaders for field trials to demonstrate how an 8-week supplement of 2 grams of algae a day can end micronutrient malnutrition. This natural biosolution for micronutrient deficiencies works for adults and for children. Pregnant mothers with insufficient healthy food too often have premature or stillborn births.

Some infants experience stunting or wasting due to lack of nutrients during womb life. These unfortunate infants are unlikely to ever experience a normal, productive life.

Protein-energy malnutrition, (PEM) is another threat. If fetus fails to receive good nutrition during womb-life, major organs fail to grow properly. This results in stunting, slowing of linear growth and incomplete growth of major organs.[75]

Behavioral changes occur such as irritability, apathy, decreased social responsiveness, anxiety, and attention deficits that lead to problems in school. Early childhood effects of micronutrient or PEM malnutrition may be irreversible.[76]

Freedom foods provide pregnant mothers and their fetuses with the nutrients they need for full-term, healthy births. These nutrient-packed foods ensure newborns avoid developmental disabilities so they can achieve their full potential.

Urban children may be taller than their rural colleagues due to nutritional differences. SCAD plans field demonstrations guided by National medical leaders to show how improving diets for rural mothers and children eliminate stunting, wasting, premature births and developmental disabilities.

## *Obesity and diabetes*

Many societies experience an epidemic of obesity and diabetes. Fossil foods are commonly made from low-nutrient food grains such as rice, wheat or maize. These hard, starchy and bitter grains are not palatable raw. Grains undergo extensive processing to make them palatable by adding extra fats, sugar and salt. Processed foods deliver empty calories – calories empty of nutrients.

Freedom foods provide tasty, high nutrient density nourishment. Algae are naturally soft with a sweet, neutral taste. Freedom foods are naturally low in calories, unhealthy fats, sugar and salt. Fossil foods drive the nosh response in the brain demanding more food. Freedom food provide a satiate brain signal that squashes nosh. It overrides nosh with a feeling of fullness.

Ecolanda creates goals for decarbonization, wastewater remediation and degraded cropland restoration. Each Ecolanda site creates goals for bioproduct creation in concert with the National Team.

The graphic provides examples of bioproducts sorted by value and volume.

### High value, low volume

- Novel biochemimstry molecules
- Sensors — medical & industrial
- Clean water, air and ecosystems
- Nutraceuticals and cosmeceuticals
- Biodegradable textiles and materials
- Organic LEDs, wiring and structures
- Malleable and insulated electronics
- Photovoltaics, recyclable electronics
- Battery membranes and electrolytes

### Low value, high volume

- Emulsifiers, enzymes & preservatives
- Hygiene and absorbent bioproducts
- Construction materials and biofoam
- Aerospace structures and interiors
- Aerogels, coverings and insulators
- Biopaints, pigments and coatings

### All SCAD Ecolanda bioproducts are:
- Biodegradable
- Net-zero waste and pollution
- Healthy for their intended use

### Medium value

- Human food - functional and ingredients
- Plant-based meats and hybrid foods
- Superior nutrition, micronutrients, vitamins
- Specialty construction elements and fibres
- Automotive parts and furniture
- Packaging, films, coatings and bioplastics
- Paper, packaging, fillers and insulation
- Biofuels: oil, diesel, ethanol and jet fuel
- Biodegradable cementa and asphalta
- Feed - meat, dairy, farm animals & fish
- Ecological recovery — BioRestore
- Biotextiles and bright pigments
- Highly productive biofertilizer

## Environmental health

SCAD make commitments in every Ecolanda project to improve environmental health. Similar to human health, needs vary by nation.

Most areas have need for cleaning and restoring health to air, water and ecosystems. Each country makes a unique set of environmental goals.

## Heavy metals

Many communities rely on deeper and deeper wells for drinking water. Deep wells increase the likelihood of contamination from deadly heavy metals such as lead, mercury, cadmium and arsenic. Each of these toxins can cause severe damage to the developing brains, hearts and other major organs especially in children. The poisonous metals destroy adult health too.

Ecolanda biosystems can remove toxic heavy metals from water so they can do no harm. Heavy metal biomass may be made into bioplastics, bioresins and bioconstruction materials where they stay inert and non-threatening for decades.

BioRenew can remove toxic heavy metals from drinking water. What about children and adults that suffer from brain and central nervous system dysfunction as a result of ingesting heavy metals?

An algae health food, Spirulina, allows the body to assimilate the tiny algae cells that chelate with heavy metals. The cells can pass through the blood/brain barrier. The body sluffs them off and passes them out of the body in the urine. In regions plagued with heavy metals, the ability of freedom foods to detox heavy metal poisoning can save millions from painful and ugly disability or death.[77]

## Land reclamation

Large tracks of crop land have been abandoned due to erosion, salt, compaction and exhaustion. Expanding deserts, deforestation and invasive weeds threaten ruin of other lands. Rising oceans take a harsh toll on some of the most fertile crop lands, coastlines and river deltas.

SCAD creates plans and timelines to repair these lands and prepare them for restoration.

## Coral reefs

Nearly half of the world's population live within 100 km, (60 miles) from a coastline.[78] In 1990, there were 10 megacities, with over 10 million people. The UN anticipates 41 megacities by 2030, with 36 on coastlines.[79]

Sea levels have risen an average of 40 cm, (1.3 ft), over the past 100 years. NOAA's models predict the rate of sea level rise to increase – substantially.[80] People living along coastlines are highly vulnerable to sea level rise.

An international study revealed that coral reefs protect billions of people along coastlines and river deltas from rising sea levels and damaging wave action.[81] *Nature Communications* reported that **coral reefs reduce wave energy by 97%** and **wave height 84%.**[82]

Coral reefs, rainforests of the sea, are home to the most biodiverse and productive ecosystems on earth. Corals are similar to fungi's structure in lichens on land. Corals and lichens depend on their symbionts, algae, for 90% of their nutrients and 100% of their pigments, color.

Coral reefs have a global economic value of over $375 billion a year. Reefs provide food and resources for more than 500 million people in over 100 countries. Unfortunately, coral reefs are dying, primarily as a result of industrial pollution. Nearly 90% of corals are under threat of extinction by 2030.

SCAD will work with global biotechnology experts to develop algae cultivars that can survive and thrive in warmer and more polluted oceans. These robust algae can be introduced to corals in order to improve coral survival.

## *Peace microfarm biosolutions*

Peace microfarms preserve natural resources by biocycling waste streams, which may avoid conflict or war over land, water, fuel or fertilizer. A peace microfarm can clean wastewater and provide all the essential nutrients to avoid malnutrition for a community.[83]

Local microfarms can eliminate malnutrition and micronutrient deficiencies, in rural and urban areas. A single 50 m$^2$ microfarm can deliver enough Spirulina to cure **1,350 children** and/or pregnant mothers from the curse of premature birth due to malnutrition.[84]

Example: 50 m$^2$ (544 ft$^2$) microfarm, (surface area), About 3m by 17m. Volume about 15,142 L (4,000 gallons) at 15 cm depth. Microfarm yields about 135 kg of Spirulina a year.

Antenna research shows that an 8-week Spirulina treatment with 100 g (total) resolves child malnutrition. Microfarms grow microcrops that produce healthy protein 30 to 50 times faster than field crops such as food grains. A microfarmer can harvest about 30% of the algae biomass daily or choose to harvest a higher percentage every two days during sunny weather. Growers can produce algae food and bioproducts year-round in many regions.

Each microfarm may employ several people. Microfarms will not make them rich but will provide the means for healthy food for their family and food to sell in their community. An estimated 100 small spirulina producers are growing food locally in French spirulina microfarms as far north as Normandy.[85]

Robert Henrikson tends an algae microfarm of his design. Robert created an excellent "Getting started checklist" for creative people interested in building a microfarm.[86]

*Peace Microfarms: A green Algae Strategy to Prevent War* explains how algae microfarms give growers the freedom to produce food, feed and other valuable bioproducts locally.[87] Peace microfarms avoid war by producing food and other forms of energy with minimal fossil resources. Resources saved eliminates the need to fight over scarce food production resources.

Microfarms can provide social justice, since only modest physical labor is required and there is no dust, pesticides, heavy machinery or poisons.

Women, physically handicapped and elderly people can grow highly nutritious food locally. Some microfarms have been designed for Wounded Warriors who have lost limbs in war.[88]

Families can support themselves and stay in the communities they love.
A microfarmer may cultivate one of many algae species. She may choose to dry the product or sell it fresh locally. Fresh or frozen algae may be eaten directly with no processing or cooking.

Peace microfarms can produce healthy food independent of altitude, latitude, climate or geography. Production systems can be sited on empty lots, rooftops, urban gardens, rail rights-of-way, and balconies. Microfarmers can grow enough food in cities to feed the entire city.

Microfarms will create good jobs for families that live in inner-cities, slums, food deserts and rural communities. Urban microfarms can significantly reduce transportation costs and black soot pollution from diesel truck engines because local food production nearly zeros out transportation. Microfarms can biocycle nutrients and grow excellent biofertilizer and animal feed for urban farmers and family animals.

SCAD will create public-private partnerships to build microfarms and support growers.

# 16. SCAD Management

*No matter how complex or affluent, human societies are nothing but subsystems of the biosphere, the Earth's thin veneer of life, which is ultimately run by bacteria, fungi and green plants.*　　　*– Vaclav Smil*

*One touch of nature makes the whole world kin.*　　　*– William Shakespeare*

SCAD learns biological solutions from nature and uses abundance methods that produce sustainable food that is healthier for people, animals, plants and our planet.

Global hunger needs to be addressed as a nutrition crisis. Solutions must provide people with affordable, high-quality protein and all the essential macro and micronutrients they need to gain and sustain health and vitality.

SCAD operates as a virtual company and has commitments from experts with 20 to 40 years of experience in each key production and restoration system. SCAD commits to hiring a majority of host-country people who are or will be trained in SCAD abundance technologies.

SCAD staffs like to a Hollywood movie studio where key people are brought together when needed to produce each film.　SCAD verticals include Agri-Energy-Water-Waste-Eco-City with initial staffing shown approximately.

*SCAD initial staffing:*

- Permanent – Core executives only
- Project staff 10+ leaders per vertical
- Project consultants — 10+ per vertical
- Plan, Design, Build - 100 global experts
- R&D scientists - 300+ global experts
- Technology experts – 300+
- Operations – 9,000+
- Training & educ. – 100+ at all levels
- Quality & Safety – across verticals
- Employment equity – commitment

SCAD has commitments from over 100 people who are operational experts in their disciplines. These world-class leaders include Ecolanda staff, architects, engineers, consultants and Post-Docs.

SCAD's virtual firm organizes around the five vertical sectors. SCAD leaders and consultants assist with design, planning, systems integration and building effective biosystem infrastructure.

| Smart Agriculture | Smart Energy | Smart Water | Smart Waste | Smart Ecocity |
|---|---|---|---|---|
| **Waste by type** | **Total energy used** | **Fresh water used** | **Nutrients recovered** | **Energy efficiency** |
| • Waste recovered | • Total energy made | • Alternatives used | • GHG by type – used and lost | • Eco efficiencies |
| • Gas, liquid, biosolid | • Solar, wind, hydro | • Brine, waste, ocean, recycled | • Waste collection Waste efficiencies | • Resident sat |
| • Water efficiency | • Tides, geothermal | • Water efficiency | • Phosphorus, carbon, others | • Green transport |
| • Feed efficiencies | • Efficiency metrics | • Clean water | | • Health, happiness |
| • Productivity | • Storage efficiency | | | • Sustainability |

Some leaders will work only through the design and build phase. Many will stay engaged to assist with operations.

Leading biotechnology scientists have agreed to engage with Ecolanda. These are top scientists and engineers who benefit from global academic and business networks. SCAD has commitments from several dozen top algae whisperers who have extensive experience in operations.

SCAD has assembled an Advisory Board of top biotech and environmental scientists. They will guide SCAD and provide talent – post docs, graduate students and their graduate network – to address specific opportunities and R&D.

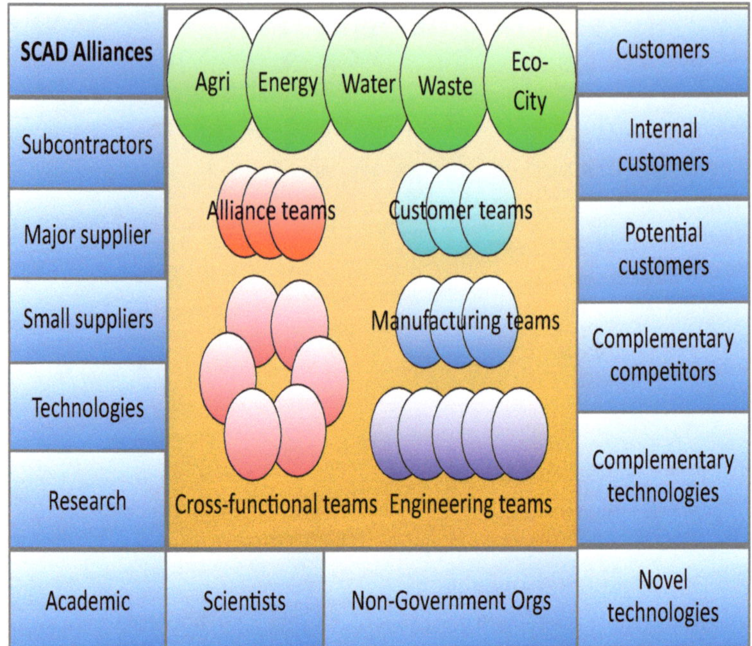


SCAD's virtual organization design offers substantial advantages:

- Larger talent pool. SCAD employs the best people globally, with a majority of the 10,000-person talent sourced from the host country.

- Lower overhead and operating costs. Virtual organizations enjoy significant savings in operational costs as the company pays only for work engagement.

- Higher employee efficiency. Employees get more work done without overbearing office transactions, which allows them to accomplish more in less time.

- Higher employee satisfaction. Employees are happier when they are relatively independent and report lower stress levels and higher morale.

- Higher employee retention. Employees are strongly invested in the great work they do for their country and our world in Ecolanda.

- Access to advanced technologies. SCAD hires people with strong skill sets and continually refreshes skills and abilities.

SCAD has strong commitment to social equity in hiring. The highly diverse staff will represent many countries, but a majority of the permanent staff come from the host country.

The Ecolanda smart ecocity architecture and organizational plans are designed to attract and retain top talent globally and regionally.

SCAD operates based on science which creates big data and advanced metrics.

## *Metrics*

Advanced metrics make up the SCAD Ecolanda backbone. Advanced metrics include Internet of Things, IOT, big data, advanced analytics, machine learning and AI.

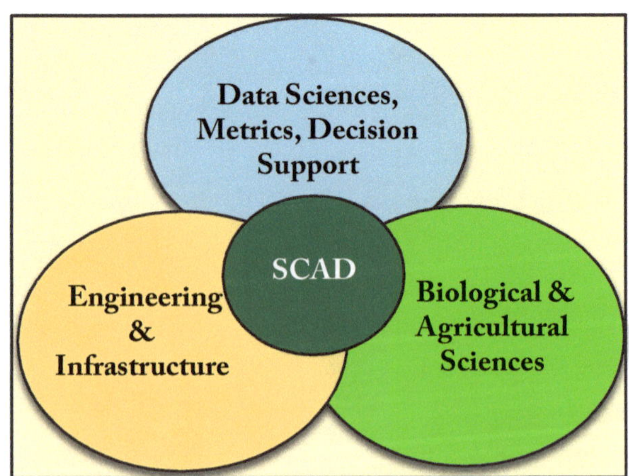

SCAD succeeds with multiple complex advanced biotechnology and renewable energy systems by relying an array of sensors, monitors and metrics. The table provides examples.

### SCAD Ecolanda
#### Agri-Energy-Waste-Water Bioechnologies

| Bioremediation | BioRenew | Facilities |
|---|---|---|
| **Air**<br>• CCU<br>• Piping<br>• Concentration<br>• Combustion<br>• Algae biosystems | **Bioproducts**<br>• Mass balance calculations<br>• Extraction technologies<br>• Compound analytics<br>• Directed genetics<br>• Bioprospecting | **Biocycling**<br>• Nutrient capture<br>• Monitors<br>• Controllers<br>• Analytics & metrics<br>• Novel materials |
| **Biosolids**<br>• Biological analysis<br>• Sources & methods<br>• Contaminants<br>• Non-biological<br>• Trace compounds | **Advanced compounds**<br>• Search and find<br>• Secondary research<br>• Partner labs<br>• Historical medicines<br>• Green chemicals | **Platforms**<br>• CCU<br>• Bioproduct options<br>• Optimization<br>• Cultivate & harvest<br>• Bioprocessing |
| **Water**<br>• Biological tests<br>• Salinity, pH<br>• Poisons, heavy metals<br>• Velocity & volume<br>• Pharmaceuticals | **New opportunities**<br>• Nutraceuticals<br>• Cosmeceuticals<br>• Gut microbiota<br>• Health & medicines<br>• Brain science & meds | **Integration**<br>• Carbon sources<br>• Nutrient sources<br>• Logistics & monitors<br>• Resource accounting<br>• Energy and water |

Ecolanda metrics provide targets and accountability for each area of bioremediation, Bio-Renew and operational facilities. Cultivation and renewable energy generation use automated systems to monitor carbon capture, nutrient recovery, reuse and growing system parameters 24 /7.

Monitor and production data are shared across Ecolanda. These data drive internal accounting and finance. Targeted data is shared with SCAD-affiliated universities, agri-institutes and farmers to support training, R&D and knowledge transfer.

## Revenue model

The SCAD Ecolanda architecture relies on systems integration. Inputs, internal processes and outputs for each process are continuously measured and monitored by smart systems.

Many bioproducts such as bioenergy, biofertiliser and biofeed supply consumers in other areas within Ecolanda. For example, Nrich supplies feed for aquaculture, animals, dairy and nutrients for CEA horticulture.

The flow model below illustrates how Nrich transforms waste streams to revenue.

SCAD invented a smart tool to capture relevant data across these integrated processes. ROSIE allows SCAD to account for natural resources.

ROSIE, Return On Smart Innovation and Environment, continually monitors reliability and validity for each Ecolanda sector.

ROSIE allows leaders, managers and operators to track natural resource avoidance, savings, substitution and efficiencies.

---

**Nrich transforms waste streams to revenue.**

- BioRenew uses algae and photosynthesis to recover, recycle and reuse nutrients.
- BioRenew remediates waste streams into carbon and other valuable nutrients.
- Sensors measure and track all inputs, process steps and bioproduct outputs.
- Nrich monetizes remediation, carbon capture, bioproducts and greenwill.

**Nrich BioRenew revenue model = RW + CC + BP + GW**

RW (Remediated waste) + CC (carbon capture) + BP (Bioproducts) + GW (Greenwill)

| RW | CC | BP | Greenwill |
|---|---|---|---|
| • Polluted air | • Carbon | • Feed / food | • Eco goodwill |
| • Wastewater | • Poisons | • Fertilizer | • Eco metrics |
| • Botanicals | • Nutrients | • Nutraceuticals | • Competitive advantage |
| • Mine powder | • Heavy metals | • Ingredients | |
| • Harbour waste | • Trace elements | • Biomaterials | • Teaching |

### Risk management

**SCAD Risk Management**

ROSIE and the IoT verticals drive risk management. These tools populate data and information 24/7.

Sensors and monitors drive multi-function dashboards that display information to the communication centre. Most dashboards are available remotely via phone or PC.

The IoT verticals create massive amounts of data that are assembled in templates that make them easy to comprehend.

SCAD shares extensive data on animal and plant cultivation systems with colleges, universities in

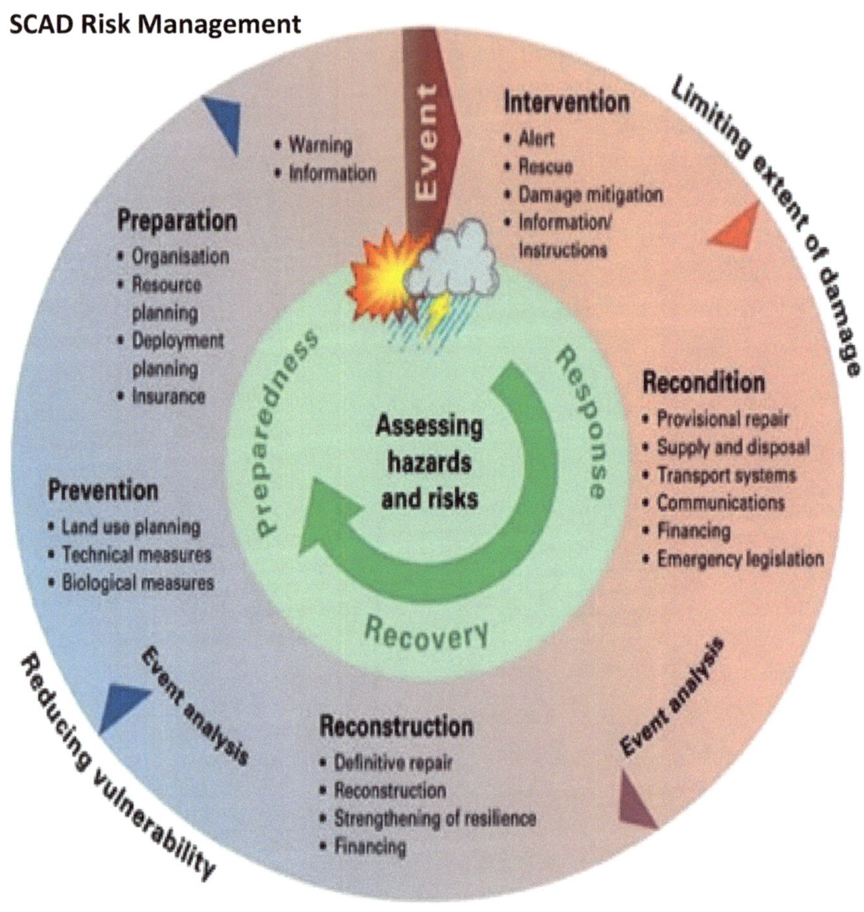


agricultural institutes nationally and with affiliated institutions globally. These strong agricultural relationships benefit agri-leaders with superb data for their research and publications. The subject knowledge experts benefit Ecolanda by providing checks and balances as well as credible validation for developing technologies.

SCAD employs a world class risk management team than anticipate and mitigate risks. A major risk, contamination, receives special attention. Ecolanda design places agri-production in clean-room facilities. Associates dress in fresh gear every time they enter or leave a cultivation facility.

Production operations include extensive automation that tracks key growing parameters to assure consistent, healthy cultivation stays up to high standards. Similar metrics track produce

quality on relevant cultivation attributes, especially growth efficiency and nutralence.

Risk managers continually assess obvious and obscure threats. Checklists are read and followed for expected and unexpected events. Safety drills are common to ensure all associates understand how to minimize risk.

When crisis events occur, SCAD follows a rigorous reconstruction process to identify issues and build resilient responses.

Ecolanda engages many novel biotechnologies. Most are designed from the foundation with safety and risk management in mind.

Strong prior preparation avoids emergencies. SCAD is committed to rigorous preparation and planning.

### Triple bottom line

Sustainable systems, also called triple bottom line, drive decisions across the Ecolanda verticals.

Social development includes far more than job creation to ensure people equal opportunity for jobs, training and projects.

SCAD will have a strong community outreach process that builds local engagement and economic impact.

SCAD has created the architecture for Ecolanda to be the highest productivity agri-energy centre in the world. Economic development works closely aligned environmental stewardship.

Environmental emphasis begins with processes that ensure health and safety for all associates and the community.

Sustainable production focuses on minimizing the use of extracted resources. Resource efficiency, like energy efficiency drives cultivation decisions. SCAD not only plans to prevent pollution but also to use a suite of proprietary, licensed and public biological tools to restore degraded ecosystems. SCAD has a strong commitment to replenish lost biodiversity. Environmental restoration provides the foundation for biodiversity recovery.

### Summary

Ecolanda will lift citizens' quality of life with 10,000 good jobs and over 50,000 support jobs. People will benefit from healthy food and affordable housing. Comprehensive education and training will continually improve living standards and economic development.

Economic growth will occur while producing healthier and more affordable food and green energy with zero waste or pollution. Ecolanda plans to cultivate enough food for 5 million people.

SCAD produces 90% of needed biofertiliser internally for all forms of food, feed and fibre cultivation. SCAD produces 80% of needed animal feed internally. Animals production. includes a million fish a year, plus other meat and dairy animals. The emerald bioeconomy will demonstrate global leadership in addressing climate change by capturing and reusing over 70 million metrics tons of $CO_2$ annually.

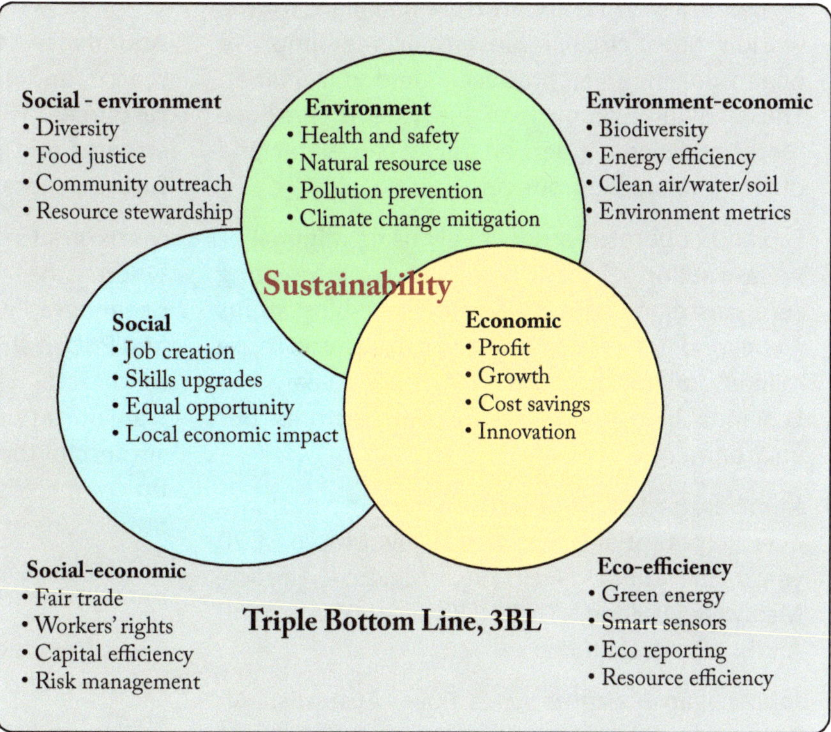


Ecolanda will clean and restore millions of liters of blue water, which benefits the community. Nrich and BioRenew will remediate and restore thousands of hectares of degraded land and ecosystems.

### What defines success?

Ecolanda succeeds when citizens and associates agree that together we have resolved two of the greatest challenges of our time to build:

1. Robust economic and social growth and development while restoring our environment.
2. A successful 7-generation emerald bioeconomy.

### *John G. O'Hare, Managing Director SCAD Queensland, Hong Kong, Malaysia & India*

**John plans, designs and leads the buildout and operations for all SCAD Ecolanda projects.** He has responsibility for site selection, capital raising and building the SCAD leadership team.

Each SCAD Ecolanda™ benefits from John's 40 years of R&D in sustainable agri-energy systems.

He created the first architecture and plans for the world's finest circular biosystems that improve health for people, producers and our planet. These highly productive biosystems produce food, medicines, green energy, clean water and other valuable bioproducts

Ecolanda operates productively using minimal or no extraction of fossil resources while creating zero waste and zero pollution. These biosystems go beyond net-zero carbon and capture and reuse millions of metrics tons of carbon a year. The Ecolanda bioeconomy cleans and restores our environment.

John has used his site-selection matrix to evaluate over 50 potential Ecolanda sites over the past 20 years in China, Australia, Mexico, Greece, Malaysia, Ireland, Great Britain, Ghana and Andhra Pradesh, India.

John began his career as a Royal Australian Air Force Squadron during the Vietnam conflict. He followed that experience working for 7 years for Qantas Airways. He farmed and developed a series of nurseries across Australia and the Asia-Pacific region.

John worked over a decade as a registered equity derivatives trader in Australia and the Chicago Board of Trade. He holds a degree in Business from the University Southern Queensland. He earned several credentials for options trading in Australia and retail and institution certifications on the Chicago Board Options Exchange.

John invests considerable time investigating renewable energy methods that maximise carbon capture and conversion to Emerald Energy that is completely clean.

### *Mark R. Edwards, Director, Biotech and Ecology*

**Mark pursues 7-generation stewardship.** This involves global R&D on circular biotechnologies that drive sustainable bioeconomies for 140 years.

Ecolanda offers opportunities for stewardship with nations. We jointly design, build and operate abundance biosystems that biocycle carbon and other nutrients to eliminate the need for extracted fossil resources, waste or pollution.

Abundance methods avoid the need for war over scarce and increasingly expensive fossil natural resources. The BioRenew process biocycles waste streams and preserves precious resources for our future generations.

Mark graduated from the U.S. Naval Academy where he studied mechanical engineering, oceanography and meteorology. He earned an MBA and PhD in strategic marketing at ASU. He served as a director for the largest global food and transportation company and assisted in a series of successful food, agribusiness and technology start-up companies. He served as a professor at Arizona State University for 39 years, teaching food marketing, agribusiness, leadership, innovation, entrepreneurship and sustainability.

Mark founded and operated as CEO of a successful software and assessment firm, TEAMS Intl., for 22 years. He served as principal consultant and created advanced metrics for over 600 firms globally, including 7 of the 10 *Fortune Most Admired Companies*. He invented many innovative and award-winning metrics, including 360° Feedback. TEAMS Intl, the leader in 360° Feedback, won the prestigious Inc. 500 award for growth, leadership and profitability. TEAMS sold to an international consulting firm in 1999.

Mark authored both a business and a science bestseller. Seven books of his 18 books in the *Green Algae Strategy* series on sustainable and affordable food, water and energy won international best science and environment book awards. Mark has published over 150 business and scientific journal articles and writes the most popular blog in the algae industry, *Algae Secrets* for *Algae Industry Magazine*.

# SCAD Ecolanda — EcoSummary
## SCAD Designs, Builds and Operates an Emerald Environment with Resilient Economic and Social Development

**SMART Systems — S**ustainable **M**icro **A**bundance and **R**egenerative **T**echnologies. SCAD's smart systems use biosolutions to improve crop productivity and health for people, producers and our shared environment.

### Smart Climate — Reverse GHG Pollution

Ecolanda employs a suite of **Smart Climate** strategies.

**Smart Agriculture** with abundance methods.
- Emerald flows carbon into bioproducts, not the atmosphere
- Capture, recycle and reuse 70 MMTons of CO2e
- CCU gas from plant and animal production

**Smart Energy** with renewable abundance methods.
- Zero GHG emissions with extensive CCU
- Local energy production, smart battery and microgrid

**Smart Waste** biocycled with renewable energy.
- Zero emissions — gas, liquids, soils or ecosystems
- Green waste-to-energy supplies syngas and green hydrogen

**Smart Water**, Nrich and BioRenew clean polluted water
- Broadacre crops use 80% less water than industrial methods
- CEA uses 120% less than industrial agriculture
- Abundance agri-methods add 20% new blue fresh water.

**Smart Transportation** — public and private electric vehicles
- Zero GHG emissions — green energy powers transport
- 80% local food production reduces transportation need

**Smart Industry** — Carbon-negative, biodegradable analogs
- Building materials, concrete, asphalt, advanced composites
- Bio-textiles, pigments, bioplastics, biofilms, packaging

### Emerald Forests — Restore Forests

Deforestation generates nearly 30% of global GHG emissions. SCAD Ecolanda uses four strategies to stop deforestation, repopulate forests and restore their spectacular biodiversity.

1. **Demonstrate and train** people in abundance methods that require neither natural resource extraction nor forests cleared for cropland.
2. **Educate, mentor and support farmers** and their families to learn and use eco-friendly biosolutions.
3. **Sponsor the Emerald Forest Initiative** with an International Competition calling for innovators to show how microcrops provide farmers with higher income and less physical and economic risk without abusing forests, cropland or ecosystems.
4. **Teach people** to cultivate clean heating and cooking oil in microfarms to replace smoky fuelwood.

SCAD invents eco-solutions. These innovations can stop global deforestation in one decade.

Ecolanda **BLINC** strategies apply Abundance methods and integrate natural circular biosystems across sectors to:

1. **Stop systemic damage** – extraction, overconsumption, waste, emissions and pollution.
2. **Repair ecosystem damage** caused by destructive industrial processes and restore biodiversity.

**Biological** systems extract nothing. They do not overconsume resources or create waste. Nature's integrated sustainable systems do not pollute air, water or soil and are regenerative. Nature's circular system works exquisitely with the cycle of life.

**Light** provides abundance growers with the photons to power photosynthesis. Biosystems are the gentle engine of choice in place of fossil fuel-driven machinery that forces crops to grow.

**Integrated** systems ensure waste from one action serves as the nutrient foundation for the next. Nrich and BioRenew biocycle waste carbon and nutrients while eliminating pollution.

**Nano** and microcrops grow 20 to 50 times faster than industrial crops. These crops can be grown as freedom foods and consumer products, free from consuming fossil resources.

**Circular** bioeconomy ensures sustainable, clean production. Bioproducts are biodegradable and emissions free.

### Smart Water Strategies

Ecolanda treats water as the soul of the earth. The lifecycle and the water cycle are one. Ecolanda growers and citizens treat water as our most precious natural resource.

1. **Save** – save water with advanced technologies and smart applications that use 80% or less water compared with industrial methods.
2. **Substitute** – save potable blue water by substituting non-potable water for food and consumer goods cultivation.
3. **Sparkle** – clean waste and other non-potable water. Recover and repurpose the nutrients and create 20% more sparkling blue water than used in Ecolanda CEA.

These three strategies drive all Ecolanda design decisions because water success requires rigorous systems integration.

### Ecolanda's 10-Year Restoration Goals

- Restore health, happiness and vitality for **5 million people.**
- Restore **300,000 hectares of Emerald Forests.**
- Capture and repurpose **10 Billion metric tons of carbon.**
- Produce **10 Billion liters of fresh blue water.**
- Restore biodiversity — **10 Billion Wings and Beaks,** especially birds, bats and butterflies.
- Restore biodiverse marine life — **10 Billion Fins and Shells.**

John O'Hare johare@scadev.net
Mark Edwards medwards@scadev.net

# References


1 http://www.friendsofgaviotas.org

2 https://e360.yale.edu/features/could-abandoned-agricultural-lands-help-save-the-planet

3 https://www.edf.org/sites/default/files/10333_Measuring_Carbon_Emissions_from_Tropical_Deforestation

4 https://www.grain.org/article/entries/5976-emissions-impossible-how-big-meat-and-dairy-are-heating-planet

5    Lal, R. Soil Erosion and Land Degradation: The Global Risks. In Soil degradation; Lal, R., Stewart, B.A., Eds.; Springer-Verlag: NY, 1990; 129–172.

6          https://www.ag.ndsu.edu/publications/environment-natural-resources/environmental-impacts-of-brine-produced-water

7 https://waterfootprint.org/media/downloads/Hoekstra-Mekonnen-2012-WaterFootprint-of-Humanity.pdf

8 https://www.who.int/news-room/fact-sheets/detail/pesticide-residues-in-food

9 http://www.fao.org/newsroom/en/news/2006/1000385/index.html

10   http://phosphorusfutures.net/the-phosphorus-challenge/the-story-of-phosphorus-8-reasons-why-we-need-to-rethink-the-management-of-phosphorus-resources-in-the-global-food-system/

11 Op cit.

12 https://jbiolres.biomedcentral.com/articles/10.1186/2241-5793-21-6

13 Loopstra, R, et al. Food insecurity and social protection in Europe: quasi-natural experiment of Europe's great recessions 2004–2012, *Preventive Medicine,* 2016, 89:44, 50.

14 Graham, Linda and Lee Wilcox. Algae. New Jersey/ Prentice Hall, 2008/ 8.

15 http://www.news.pitt.edu/sites/default/files/documents/TaboneLandis_etal.pdf

16 http://ec.europa.eu/growth/sectors/raw-materials/specific-interest/critical/ (December 2, 2018).

17 Meutzner, F., Nestler, T., Zschornak, M., Canepa, P., Gautam, S., Adams, S., et al. (2018). Computational methods for battery material identification and analysis. *Phys. Sci. Rev.* 4:20180044. doi: 10.1515/psr-2018-0044

18 https://earth.stanford.edu/news/measuring-crude-oils-carbon-footprint#gs.mf81q1

19 https://www.orientenergyreview.com/oil-and-gas/nigeria-loses-2bn-to-gas-flaring-in-3yrs/

20 https://www.eia.gov/energyexplained/energy-and-the-environment/where-greenhouse-gases-come-from.php

21 https://www.ucsusa.org/resources/coal-and-air-pollution

22 https://www.nrel.gov/docs/fy01osti/29417.pdf

23 https://www.ers.usda.gov/topics/farm-practices-management/irrigation-water-use/

24https://water.usgs.gov/edu/activity-watercontent.php

25 https://www.bakerinstitute.org/publications/EF-pub-BioFuelsWhitePaper-010510.pdf

26 https://cfpub.epa.gov/si/si_public_record_report.cfm?Lab=NRMRL&dirEntryId=244350

27 https://faculty.washington.edu/ktorii/stomata.html

28 https://concretehelper.com/concrete-facts/

29 https://www.thelancet.com/commissions/EAT

30 http://connects.catalyst.harvard.edu/Profiles/display/Person/73533

31 https://sci.waikato.ac.nz/farm/content/nutrientcycling.html

32 https://www.amazon.com/Climate-Independent-Food-Survive-Freedom/dp/1479276847

33 http://www.drugdevelopment-technology.com/projects/rimonabant/

34 https://www.nature.com/articles/ijo2008235

35 https://innoventondcts.mandela.ac.za/Microalgae-Technologies/Technologies-Coalgae

36   https://www.ipbes.net/news/media-release-worsening-worldwide-land-degradation-now-'critical'-undermining-well-being-32

37 https://www.history.com/news/6-civilizations-that-mysteriously-collapsed

38 https://www.ncbi.nlm.nih.gov/pmc/articles/PMC4058318/

39 https://www.theguardian.com/environment/2020/jun/02/football-pitch-area-tropical-rainforest-lost

40 https://www.ucsusa.org/resources/whats-driving-deforestation

41 https://www.worldwildlife.org/threats/deforestation

42 https://ourworldindata.org/food-ghg-emissions

43 https://www.nationalgeographic.org/activity/save-the-plankton-breathe-freely/

44   Hongchang, Wang, 30 December 2009.
     web.archive.org/web/20091230071928/http:/www.library.utoronto.ca/pcs/state/chinaeco/forest.htm
45 http://www.fao.org/state-of-forests/en/
46 Wye Research and Education Centre. Riparian Forest Buffer Panel, 2002.
47 Jared Diamond, Collapse: How Societies Choose to Fail or Succeed, Penguin Books, 2011.
48 https://www.saveearth.info/deforestation/
49  https://www.dropbox.com/s/6vqrbb7wqoog7qf/Tiny%20Mighty%20Anna.pdf?dl=0
50https://www.conserve-energy-future.com/GreenHouseEffect.php
51https://www.who.int/news-room/fact-sheets/detail/household-air-pollution-and-health
52 Edwards, Mark R. *Green Algae Strategy*, 2008, 84.
53 www.acam.org/blogpost/1092863/ACAM-Integrative-Medicine-Blog?tag=nutrition
54 www.ncbi.nlm.nih.gov/pmc/articles/PMC5387034/
55 https://www.ncbi.nlm.nih.gov/pubmed/12362796
56 https://academic.oup.com/jn/article/137/12/2691/4670055
57 http://www.fao.org/news/story/en/item/197623/icode/
58 apjcn.nhri.org.tw/server/apjcn/18/2/index.php
59 https://www.triphobo.com/blog/best-molecular-gastronomy-restaurants
60 https://www.ncbi.nlm.nih.gov/pmc/articles/PMC3654245/
61      https://www.downtoearth.org.in/news/environment/fashion-industry-may-use-quarter-of-world-s-carbon-
     budget-by-2050-61183
62   https://www.nrdc.org/issues/encourage-textile-manufacturers-reduce-pollution
63   Ibid.
64   https://www.textiletoday.com.bd/water-pollution-due-textile-industry/
65   *Chen M, et al. 2015. Residential Exposure to Pesticide During Childhood and Childhood Cancers: A Meta-
     Analysis. Pediatrics.*
66   Edwards, Mark R,. 2011 to 2013 DARPA and NASA selected research for papers and presentations in the Habitat and the
     Medicine tracks for the 100-Year Starship Symposium in Orlando and Houston.
67 https://www.physicsforums.com/threads/algae-producing-oxygen-in-diving-cylinder.636257/
68 https://thewaterproject.org/water-scarcity/water_stats
69 https://www.dosomething.org/us/facts/11-facts-about-global-poverty
70 http://www.who.int/nutrition/topics/ida/en/
71 https://borgenproject.org/how-many-people-die-from-hunger-each-year/
72 http://emedicine.medscape.com/article/985140-clinical#b4
51 http://emedicine.medscape.com/article/912075-overview
74 http://www.worldhunger.org/world-child-hunger-facts/
75 https://www.sciencedirect.com/topics/medicine-and.../protein-energy-malnutrition
76 https://www.ncbi.nlm.nih.gov/pmc/articles/PMC3137999/
77   http://www.algaeindustrymagazine.com/can-algae-save-children-from-heavy-metals-poisoning/
78 http://sedac.ciesin.columbia.edu/es/papers/Coastal_Zone_Pop_Method.pdf
79 https://www.weforum.org/agenda/2016/11/the-10-fastest-growing-megacities-in-the-world/
80 https://coast.noaa.gov/digitalcoast/tools/slr
81 http://www.nature.com/ncomms/2014/140513/ncomms4794/full/ncomms4794.html
82 https://www.nature.com/articles/ncomms4794
83 Gershwin ME, Belay A (2007) Spirulina in Human Nutrition and Health. CRC Press.
84 https://www.antenna.ch/fr/activites/nutrition/
85 http://www.spirulinasource.com/spirulina/spirulina-farms/la-capitelle/
86 http://www.smartmicrofarms.com/about/microfarm-business-opportunity/getting-started/
87 http://www.amazon.com/Peace-Microfarms-Green-Strategy-prevent/dp/1480141208
88 https://www.woundedwarriorproject.org/

www.ingramcontent.com/pod-product-compliance
Lightning Source LLC
Chambersburg PA
CBHW041958110726
48006CB00004B/932